RIVETED

RIVETED

THE SCIENCE OF
WHY JOKES MAKE US LAUGH,
MOVIES MAKE US CRY, AND
RELIGION MAKES US FEEL
ONE WITH THE UNIVERSE

JIM DAVIES

palgrave
macmillan

First published in 2014 by PALGRAVE MACMILLAN® in the U.S.—a
division of St. Martin's Press LLC, 175 Fifth Avenue, New York, NY
10010.

Where this book is distributed in the UK, Europe and the rest of
the world, this is by Palgrave Macmillan, a division of Macmillan
Publishers Limited, registered in England, company number 785998, of
Houndmills, Basingstoke, Hampshire RG21 6XS.

Palgrave Macmillan is the global academic imprint of the above
companies and has companies and representatives throughout the
world.

Palgrave® and Macmillan® are registered trademarks in the United
States, the United Kingdom, Europe and other countries.

ISBN: 978-1-137-27901-9

Library of Congress Cataloging-in-Publication Data

Davies, Jim.
Riveted : the science of why jokes make us laugh, movies make us cry,
and religion makes us feel one with the universe / Jim Davies.
  pages   cm   Includes index.
   ISBN 978-1-137-27901-9 (hardback)
   1. Evolutionary psychology. 2. Cognitive psychology. I. Title.
BF698.95.D38   2014
155.7—dc23

                                                      2013050467

A catalogue record of the book is available from the British Library.

Design by Letra Libre, Inc.

First edition: August 2014

10 9 8 7 6 5 4 3 2 1

Printed in the United States of America.

CONTENTS

INTRODUCTION

In my first year of graduate school I worked in a condemned building. Sitting in my office was a woman I was trying to impress. We were talking about dance music. She liked club music and techno; I liked rap. I put on an acid jazz album in the compact disc player.

"How can you dance to this?" she asked.

"How can you *not* dance to this?" I replied, and then demonstrated the irresistibility of the track. For the most part, I only want to listen to music that makes me want to dance. Sorry, John Denver.

When I was young I was fascinated with psychic powers. I read every book in the libraries of both my elementary and high school on the subject, and was convinced that people had untapped mental abilities. All these books in the nonfiction section of my school's library told me that people could move things with their minds, scry with crystal balls, and predict the future. We used only 10 percent of our brains, right? What else could that 90 percent possibly be for?

I was absolutely captivated by this idea and convinced of it, until I read Susan Blackmore's sobering *In Search of the Light: Adventures of a Parapsychologist* in college.[1] It was the first skeptical book I'd encountered and it scorched and salted the lush landscape of my paranormal beliefs. First Santa Claus and now this? Ideas can be beautiful and we don't want to let go of them even when we know they're wrong.

There are things in this world that deeply resonate with us. We seek them out. They hold our attention. They feel right. I want to dance to hip-hop. I feel moved by sad, uplifting stories. I want to believe that people can move things just by willing it to happen.

You are struck by a beautiful view from a mountain cabin. You hear that everyone gets the afterlife that they imagined they'd have, and the idea is so beautiful, and feels so right, that you smile in spite of yourself. You hear a story of some terrible thing that happened to a child that gives you chills and haunts you for days. You find yourself glued to the screen, watching a close basketball game. You hear a great joke and can't wait to tell it to your friends.

With the huge variety of things we find compelling, it seems natural that a huge variety of qualities would make them compelling. There can't be anything similar about what's good about a pop song on the radio and what's moving when someone recounts their near-death experience, can there?

Yes, there can. Strange as it might seem, compelling things share *many* similarities. My purpose in this book is to tie together research from many fields. I'll do something that has never been done before and show how all these phenomena can be explained with the foundations of *compellingness*. I will show you that, like art and other sensory experiences, beliefs and explanations have aesthetic qualities that make us more or less likely to believe them. The same qualities appear again and again in riveting things, be they jokes, paintings, quotations, paranormal beliefs, religions, sports, video games, news, music, or gossip. The qualities that are common to all these things fit like a key in a lock with our psychological proclivities. I call it the *compellingness foundations theory*.

*   *   *

Understanding compellingness and how it works requires some understanding of our brains and how they were shaped by evolution. Our brains are a mix of old and newer processes that evolved at different times. They sometimes "disagree" on the meaning, importance, and value of things, and often we are clueless as to how we got our opinions. Often we are attracted to something or repelled by it and don't know why, and the reasons we dredge up are confabulations, mere guesses about our underlying psychologies.

The old brain is evolutionarily older. It's located near the top of the brain stem and the back of the head. We share much of its anatomy with other animals. It's a Rube Goldberg contraption, with special rules for this and not for that, all evolved, rather haphazardly, to help us survive and reproduce. It consists of a hodgepodge of specialized systems.

In the front of our head is the new brain, which is a general-purpose learning and reasoning machine and a system that tries to control the impulses of the older brain. It's a slow, deliberate planner and imaginer. Jazz improvisers quiet this part of their brain before performing.[2] This part of the brain is not built to do specific jobs; rather it's built to *learn* to do, well, just about anything. Where the old brain looks different depending on where you look, the new brain (particularly the cerebral cortex), looks remarkably similar no matter where you look. We have an old brain for the same reasons all animals do. We have the new brain because our ancestors got into an intelligence arms race with each other.

Because the old and new brains think with different rules, care about different things, and might even use different stores of knowledge, they often come up with different evaluations of the same situations. For example, there is a famous moral-reasoning experiment run by psychologist Joshua Greene that asks whether or not it is morally acceptable to pull a switch that will cause a train to kill one person rather than five (this is a version of a problem first proposed by Philippa Foot in 1967). Most people answer yes, such an action is acceptable, which indicates relatively high activation in the newer, more frontal areas. More emotionally salient problems, such as a version of the same problem that would require the *pushing* of a single person onto a track to save five people, show activation in the emotional, older parts of the brain. In this kind of scenario, where there is direct physical contact involved, people often report that doing so is morally unacceptable.[3]

When the new brain pulls in the opposite direction from the old, you can literally be of two minds about something. For example, your new brain can know that prepackaged cupcakes are unhealthful, but your old brain can be quite insistent that they should be devoured. Many of them. With a cold glass of milk, please. The old brain "knows" that sugar and fat are scarce and should always be eaten when the opportunity arises. Thousands of years of evolution taught it that. It doesn't know that fat, sugar, and salt are now plentiful and contributing to an obesity problem in the industrialized world. In contrast, the new brain knows that too much sugar isn't good for you. But who are you going to listen to? (While we're asking, who are "you?")

The new brain knows things because it learned them. In this case, each one of us has to learn things that run counter to what the old brain knows from evolution. Such is the source of many internal mental conflicts. The old and new have a tug of war. The new brain thinks more logically, in a step-by-step way. The old brain is not deliberative; it is intuitive. Sometimes we can get a strong, immediate sense that something is immoral. This is the old brain's influence. Then, when asked to explain it, we must engage our new brains, which struggle, often unsuccessfully, to apply what we "believe" about morality to justify the feeling. This is what moral psychologist Jonathan Haidt called "moral dumbfounding." Similarly, when choosing an immediate versus a delayed reward, the emotional part of the brain—the old brain—is active when thinking about the immediate reward, as are the frontal areas—the new brain—for the delayed reward. It could be that because we have some control over the activity of our frontal areas, we can use it to override the emotional areas when we need to. The everyday term for this would be "resisting temptation."

Because the old brain is better at context and the new brain better at rules, there is evidence that people with low working memory (people who can't remember long strings of numbers, for example) are actually *better* at some tasks that require old-brain function, particularly complex categorization tasks, as shown in a study by psychologist Maci DeCaro.[4] The new brain uses its laser focus to attend to only a few attributes of a situation. The old brain, in contrast, has a more diffuse focus, giving lots of information more or less equal weight.

We often personally identify with the workings of our new brain. When we see a scary monster depicted in a movie, we say we don't "believe" that the monster is there. However, as philosopher Tamar Szabó Gendler points out, some part of our brains must believe it's there, or we wouldn't get scared at all.[5]

What does this have to do with compellingness? Often the evaluations by our old brain make us prefer certain ideas and experiences. In general, its processes are black boxes, inaccessible to consciousness.

Intuition can feel like a burst of insight that comes from nowhere. What is really happening is that unconscious processes generate judgments and feelings that bubble up to consciousness fully formed. For example, you might get a feeling of danger or trust. You are aware of their outputs, but you can't look inside yourself to see how they work. If your mind is like an ocean, your conscious mind is just the surface of the water. Because you don't know what causes these feelings in your own head, it can feel like divine intervention or psychic ability.

We call these unconscious convictions and feelings *intuition*.

Should you trust your intuition? These unconscious processes either evolved or were learned for some reason or another, and you often cannot tell in the moment whether those reasons apply to your current situation. If you justify the feeling with reasons, those reasons are a confabulation that your conscious mind has invented.

It is not clear what the relationship is between trusting your intuition and belief in the paranormal and in conspiracy theories. However, a study by psychologist Matthew Boden showed that people who trust their hunches were more likely to have "ideas of reference," which are beliefs that things in the world relate directly to them—such as that someone got on the elevator because they were in it or that the raindrops are trying to send them a message.[6] Ideas of reference are common symptoms of mental illnesses such as mania and schizophrenia.

* * *

People have an understandable desire to comprehend the world around them. At its best, news holds people, governments, and other groups accountable for trespasses, and it is a purveyor of important information. At its worst, news plays to our hopes and fears, terrifying us without justification. Psychologists Hank Davis and S. Lyndsay McLeod conducted a large study that asked people to sort news stories published between 1700 and 2001 into classifications of their own choosing. The 12 categories that emerged corresponded to things we *have evolved to find important,* such as reputation, treatment of offspring, good deeds, violence, and sexual assault.[7] News has always been sensational. As terrorism expert Scott Atran says, "Media and publicity are the oxygen of terrorism. Without them, it would die."[8] Reports of terrorism increase fear and anger, and lessen feelings of forgiveness, found a study by psychologist Alice Healy. Watching nonfiction crime documentaries makes people think the national crime rate is increasing and makes them less confident in the criminal justice system. People who watch less TV are more accurate judges of risks.[9] Unfortunately, we are most likely to remember the least likely events.

What can our desire to know things about the world, and about people, tell us about our interest in fiction?

Finding Nemo is an animated movie about a father clownfish trying to find his son who gets lost. I remember marveling at the fact that this film brought tears to my eyes within five minutes. Let's examine the drastic differences between watching this film and witnessing a real-life tragic event. First, one is watching a *film,* not actual, physical people. Second, it's about filmed *fish,* not filmed people. Third, the fish are animated by computer programs. That is, it is not a film recording of real fish. Fourth, it's not even *about* real fish—the characters are completely fictional. Reflect for a moment on the absurdity of somebody (me) crying over a screen representing fictional fish spawned by computer graphics.

Long ago, our ancestors saw very few things that were not what they appeared to be. Some notable exceptions would be mirages, reflections of objects in liquid, dreams, echoes, and how a stick appears to bend when placed in water. But most of the time, if you saw a person, there was actually a person there. A sense's "proper domain" is that which it was designed (by evolution) to understand. But the "actual domain" is anything that triggers it. In contrast, the modern world constantly exposes us to images and sounds depicting things that are not right in front of us. We are exposed to people, but also representations of people that we substitute for the real thing. When we see someone on television say she likes ice cream, we believe that the actual person represented likes ice cream, not merely her televised image. We hear voices of people who are not present on the radio, on podcasts, and on the phone. We see events and people animated on screens. We see photos in magazines and on billboards. Of course the photo is real, but the image of the person in the photo is not. It might be that fully seeing nonreal things as they truly are—for example, seeing a picture of a woman only as a piece of paper and ink—is, in the long run, evolutionarily adaptive, just as it *might* be worth a designer's time to try to get the grocery store door not to open for dogs. The door was designed to open for people (the proper domain) but opens for dogs too (part of the actual domain). Environments, however, be they cultural, technological, or natural, can change so fast that evolution can't keep up. Photography was developed so recently that there simply has not been enough time for evolution to make the older parts of our mind good at distinguishing, say, actual people from pictures of people.

This contrast between what you know is true and what you perceive has a similar effect in optical illusions. We experience one thing, but in some sense "know" that the experience isn't real. Consider this effect for a moment, because its meaning is profound and its ramifications great. Although we all have experience with optical illusions, it's important to appreciate that we have no control over their effects on us. We "know" that the lines are the same length, but we still perceive them to be different. Optical illusions are designed to put your brain in this befuddled state of affairs. Your early visual system is being tricked and no amount of convincing from your deliberate thinking processes can make the visual system interpret it otherwise.

A similar effect is occurring whenever you perceive representations in any art form. Your old brain (and your perceptual areas) believes that what you're experiencing is real and often your new brain knows better at the same time.

Films are multimedia experiences that at least appear to depict real people, in that the light patterns that play on our retinas resemble the patterns that we'd get if we were looking at real people. But literature is even more mysterious in that we can become fully engaged when we are being exposed *merely to words*. Language-based narrative arts have an even greater chance to capture our attention than visual arts, because language is devastatingly effective at communicating the nuances of social changes, mental states, and relationships. Here are six words (sometimes attributed to Hemingway) that I'm sure will stir the emotions of a good percentage of readers: *For sale. Baby shoes. Never worn.* How can six words pack such a punch? Philosophers call our emotional response to fictional characters the "paradox of fiction." Why would the part of our minds that wants to understand our actual peers get excited when reading a book or looking at what is obviously a static painting?

We'll begin by looking at why people engage in oral storytelling, the likely precursor to literature. One of the fabulous things about language is that it allows us to communicate the lessons we've learned. Without communication, learning can only occur through direct personal experience or direct observation of others. Unfortunately, some important lessons, such as the dangers of wandering into enemy territory or the deadliness of getting caught in front of an avalanche, are costly to learn by trial and error, as the learner is likely to die during or shortly after the event they would be learning from. However, with language, a lesson learned by a single individual can be communicated to a great number of people. Not only can a lone surviving individual tell her tale, but others can tell it at second and third hand. Lessons can be communicated throughout an entire culture, sometimes lasting generations. It is likely that the communication of wisdom was the original function of storytelling.

As philosopher of art Denis Dutton eloquently put it, "stories provide a low-cost, low-risk surrogate experience."[10]

Modern humans can tell and appreciate fictional stories, be they crafted entertainment or outright lies. However, we still learn from and take to heart these made-up stories because we've evolved, culturally or genetically, to do so, often without even intending to. Psychologists Elizabeth Marsh and Lisa Fazio ran a study that found that people do not spontaneously keep an eye out for falsehoods.[11]

If we don't know whether a story we hear is truth or fiction, or some combination thereof, perhaps we will believe some part of it, just in case it contains some useful information. For example, a story about a bear attack might make us wary of approaching an actual bear, whether we know the story to be true or not. What about stories that we *know* are fiction, either because they are explicitly labeled as such, or, like *The Wonderful Wizard of Oz*, because they are simply too fantastic to believe? Although each of us feels as though our own mind were a single entity, under the hood of consciousness different parts of our brain fight for their points of view like argumentative parliament members. In the case of fiction, the old brain might believe the story even if the new brain doesn't.

*   *   *

In this book I present a unified explanation of compellingness. Riveting things appeal to the foundations of compellingness. Those foundations are:

1. We are interested in people. We love to see people and learn about people, and we love explanations that involve their desires, loves, and conflicts. We prefer paintings of people and religious myths featuring personified gods and spirits.

2. We pay particular attention to things we hope or fear are true. Hope gives us pleasure, and fear, though we might not particularly enjoy it, demands our attention. This is why we are susceptible to believing in miracle cures, get-rich schemes, and attractive salespeople. It also explains the fear of hell and

why we willingly expose ourselves to terrifying things like thrill rides, horror movies, and television news.

3. We delight in finding patterns. When we notice regularity in the world, we understand the world better and, thanks to evolution, we get rewarded with a rush of pleasure. Compelling things always have patterns, be they a repeated chorus in a song or a repeated religious ritual. When we notice the same pattern again and again, however, we can get bored. We are curious beings that seek out new things to understand and new patterns to discover.

4. We are attracted to incongruity, apparent contradictions, novelty, and puzzles. When there is something askew, something we don't quite get, we are intrigued and want to figure it out. Art and religion both play on our love for incongruity and for pattern, pulling us in with something incongruous and then allowing us to feel pleasure when we discover an underlying meaning or pattern.

5. The nature of our bodies—the nature of our eyes and other sense organs, for example—affects what kinds of things draw us in.

6. We have certain psychological traits, many of which are evolved, that make us like and dislike, believe and disbelieve.

So how do these foundations explain what we find riveting?

* * *

First, the arts. I will mostly focus on folk and pop art, as I believe they better exemplify what people actually like, as opposed to fine arts, which are subject to intellectualizing by specialists and as such have the potential to be further removed from our natural and general cultural proclivities.

Why do people like to experience art (and things we experience in an artlike way) and how does art affect them?

Compelling art works because it reliably creates experiences in audience members. David Byrne reflects on creating music in this way: "Making music is like constructing a machine whose function is to dredge up emotions in performer and listener. . . . I'm beginning to think of the artist as someone who is adept at making devices that tap into our shared psychological make-up and that trigger the deeply moving parts we have in common."[12] What's important is the artistic *experience*, which happens in the mind of the audience. It's a reaction.

Some might think that understanding why people like visual art is impossible because beauty is in the eye of the beholder. But as psychologist of art Ellen Winner points out, there is considerable agreement in aesthetic preferences, even among people of different sexes, intelligence, personality traits, and culture. There are indeed differences, but our similarities are undeniable.[13]

We tend to find beautiful images that show us a world that, were we actually in it, would be a great place to be. Even compelling non-representational images are liked as a result of sharing patterns with real-world things that we prefer to be in the presence of. This covers more than just art. Seeing a beautiful vista or a sunrise can elicit the same kind of response, for the same reasons. The Dinka people, in east Africa, create almost no visual art, but they appreciate the markings on cattle.[14]

Our preferences for landscapes reflect our desires for safety, information, and resources. As we will see, certain colors produce primal, cross-cultural reactions. As social creatures we like to see other people in images. As sexual creatures, we like to see attractive people.

We find disturbing images compelling, since they tend to hint at dangerous, important information. Even if we are horrified, we can't look away, much like rubbernecking drivers on the highway who must slow down to see what happened at an accident site. We either like to see images that show the world as we would like it to be, or as we strongly would prefer it not to be. It's the middle area that's boring—the anti–sweet spot between hope and fear.

When we see visual patterns, we are delighted, because seeing a pattern is noticing a predictable regularity in the world that we might be able to exploit. Patterns are breaks in the chaos. Visual rhymes, repeated colors, symmetry, and the repetition of symbols in a painting or a series of photographs attract us.

If we see something too often, or the patterns viewed are too familiar, we get bored. We need challenges to maintain our interest—the enigmatic *Mona Lisa* smile, an unexplored territory, or some other incongruity. There is a sweet spot between pattern and incongruity. An image is boring if it's too simple or too complex.

The location of this sweet spot differs from person to person, and perhaps even from mood to mood. The more familiar one is with an art form or style, the more incongruity is allowed. Everyone can appreciate Norman Rockwell, but it takes some knowledge to appreciate Chinese calligraphy or Jackson Pollock's drip paintings. Some paintings have surface patterns that are easy to appreciate, but further investigation reveals deeper incongruities for the audience to resolve. As people grow in expertise, they can detect more features and appreciate the art more.

*   *   *

Because they involve visuals as well as words, the performing arts and film are at the crossroads of literature and the visual arts. Even more than in painting and photography, they are about people. Nearly all films and live performances involve human or humanlike characters, ones that appeal to our desire to observe people in any situation.

Like paintings and photographs, films and performances include visual motifs that are played out over time as well as in space, appealing to the "pattern" compellingness foundation. For example, several dancers might be doing the same motions at the same time or similar movements at different times. Ballet has a vocabulary of moves. A film might use a limited palette of colors, thereby increasing the familiarity.

Our love for incongruity influences a film's or a performance's appeal as well. The works range from simple to incomprehensibly complex, all to be appreciated by different audiences, in different moods, at different times. We can stand extreme incongruities for only a short period of time, allowing music videos to be more avant-garde than feature-length films. This is why absurdist full-length theater is sometimes difficult for the general public to appreciate.

* * *

The narrative arts include any art form that uses a story element, such as theater and theatrical improvisation, novels, films, contemporary (urban) legends, tabletop role-playing games, and many computer games. People also find meaning in their lives by constructing them as narratives.[15]

Our obsession with people plays out most strongly in narrative, where conflicts between characters are ubiquitous. When we learn about the lives of characters, our old brains react as though the characters were real people, and we try to learn lessons from them. In fact, reading fiction has been found to correlate with empathy and unselfish behavior. White children who read stories with African American characters had improved attitudes toward African Americans—not only more than kids reading stories about white children, but more than children who interacted with *real* African American children on a shared task!

In one study, children who read stories *without* descriptions of mental states (e.g., "Jill was happy") did better at a test of understanding mental states of other people than children who read stories *with* such descriptions. One explanation of this fascinating finding is that forcing the audience to make inferences themselves prepares them better for reasoning about other people than having conclusions handed to them.[16]

We learn from narratives—for better or worse—whether we want to or not. Unrealistic human interaction depicted in narratives can cause problems of understanding humanity. Readers of romance, for example, probably believe more strongly in the "swept away by romance" trope. Research by psychologists Marsh and Fazio found that they have negative ideas about using condoms, have used them less in the past, and plan to use them less in the future.[17] Just as we take care of what food we put into our bodies, we should take care what narratives we put into our minds.

*　*　*

There is no human society without some musical tradition.[18] Nearly all social religious rituals use music. Our reactions to music, though they differ by culture, appear to have some cross-cultural similarities, including the perception of emotional tone. The metaphors we attach to music (high pitches being happy, for example) might be cross-cultural.

Our desire for understanding, and our desire to detect patterns, is clearly manifested in music, where repetition is crucial. Repetition happens in music at multiple levels, from a constant drum beat to the recognizable elements that make up a musical style or genre. Where we often get a sense immediately if we like a painting or not, many of us have learned that music, although we sometimes get a strong reaction at first, often takes repeated listenings before we know for sure whether we like it or not. Music is meant to be heard again and again. One can acceptably love a novel that one has only read once, but we are surprised if we hear of someone speak of a song he or she likes, but has only chosen to listen to once.

*　*　*

Some sports—such as acrobatics, figure skating, cheerleading, synchronized swimming, martial arts, parkour, and competitive ballroom dancing—occupy a gray area between games and art. Though sports are not considered art, there can be no doubt that our enjoyment of sports involves aesthetic appreciation, and that some sports, such as synchronized swimming and figure skating, tread the art/sport border. But even prototypical sports, such as basketball and soccer, share similarities with an artistic performance: people take pleasure in watching it, skill is involved with the performance, critics analyze it afterward, and there is a strong emotional response.

Much like narratives, sports involve people in conflict. Sports in which two or more players are competing directly with one another (e.g., tennis and lacrosse) might be appreciated because they represent, deep in our minds, interpersonal conflict, where solitary sports (e.g., weightlifting and archery) are appreciated for the same reasons the "man versus nature" model of plotting are.

Team sports trigger our sense of loyalty, one of the foundations of moral psychology. People align themselves, in their minds, with a team, and feel joy when it succeeds and sorrow when it loses. Loyalty matters to men and to women, but men tend to be loyal to coalitions and women to two-person relationships.[19] This would predict that men prefer team sports relatively more than women do.

Our love for repetition is satisfied with the ritual aspect of the rules, such as the dropping of the puck between two hockey players or the foul shot in basketball. At the same time, sports always involve a kind of incongruity, in the form of uncertainty: even if two teams play each other again and again, no two games are ever the same. Within the restricted bounds of the rules, endless variations fascinate us. Some sports have different levels of strategy to them. American football, for example, involves much more strategy than, say, basketball, soccer, or racing. The players all stop and execute plays that are carefully practiced and memorized beforehand. Many other sports have only tactics, in-the-moment decisions based on the chaotic state of the field. American football is rewarding at the tactical level of throws and tackles, but rewards the careful observer with another layer of strategy.

The reasons we actually play sports mirror the reasons that we watch them (such as the appreciation of competition). We also play sports for the same reasons we play at all. We evolved to enjoy practice of things that resemble real-world skills, such as fighting, hiding, and escape.

It used be that sports were played more than watched, but with the advent of mass communication, most people are spectators. I predict there will be a similar shift in computer games. It might never be that people spend *more* time watching than playing computer games, but the spectator aspect of it is growing, with public computer game competitions that attract spectators starting to pop up.

*   *   *

Perhaps the most important aspect to our understanding of compellingness is its effect on what we believe. We hear ideas from people all the time, and we are constantly evaluating them and choosing, either consciously or not, how much to believe them. Much of what we believe is justified, either based on unbiased perceptions (e.g., your belief that you are currently reading a book), rational inference (you figured something out), expert testimony (trusting in scientific consensus), and so on. An examination of these rational ways to believe has been explored in other books. Unfortunately, there are a great many irrational faculties at work as well. What we've learned in our lives, and what we have evolved to find interesting, contributes to aesthetic qualities that influence the believability of ideas.

Take conspiracy theories. We want explanations for events (because our love for incongruity attracts us to puzzles to be solved) and we love the fact that a conspiracy theory provides one (appealing to our love for patterns and resolutions to those incongruities). Some conspiracy theories are true, such as the one surrounding Nixon and Watergate. My point isn't that conspiracy theories are never true, but that they have compelling qualities that should make us particularly cautious about them.

We have a bias to attend to information that supports the beliefs we already have. The theorist believes he or she has figured something out that the rest of the world doesn't understand, which provides a sense of superiority. Threats to what we believe—whether it's conspiracy theories or religious beliefs—cause us to evangelize, to try to convince others of whatever idea is under threat. If other people agree with us, the discomfort of the threatening information is thereby reduced. Evangelism results in repeated exposure in the media, which causes an availability cascade. When we hear ideas over and over, they gain plausibility through familiarity.[20]

The contemporary mythology of alien abduction is, at its heart, a conspiracy theory with a science-fiction twist. Indeed, belief in a government cover-up often goes hand in hand with the alien-visitation myth.

Aliens are person-like, appealing to our love of anthropomorphic explanations. In particular, they are persons interested in sex and violence. In addition, the alien myth has the benefit, in the minds of its adherents, of lacking the supernatural touch that makes many of those same people dismiss religion and belief in demons and spirits as ridiculous. This gives believers a sense that their belief system fits better with the scientific worldview (which, in some ways, it does).

So-called alien abductees form support groups that recirculate bad ideas the way a building with no external intake circulates stale air.

Paranormal beliefs are those that violate the fundamental and scientifically founded principles of nature. This distinguishes them from beliefs that are merely wrong, such as believing that the dog has been walked when she hasn't, or that men and women have differences in average IQ.

Using the new-brain/old-brain distinction mentioned above, many supernatural beliefs arise from biases from our old, intuitive brain. Indeed, people who use intuition more usually hold more superstitious, religious, and paranormal beliefs than others.[21]

Astrology, superstitions, the alien-abduction myth, and most religions benefit from being about people or personified forces. People at the empathic end of the autism spectrum are more prone to these beliefs, as they are more interested in other human beings.

Beliefs based on poor science play on our hopes and fears. They can promise us a just world, an afterlife of happiness, or miracle cures to our ills, and they can scare us with hidden dangers in the universe that we are too terrified to dismiss, leading to concrete consequences. For example, fears of vaccination have now manifested into a threat to public health.

Particularly in chaotic, unpredictable environments, we struggle to understand the world and come up with superstitions and false pattern matches to guide us. Once an idea takes hold, we stop looking for disconfirming evidence, and we start to see evidence of the superstition everywhere.

There is abundant evidence that, good or bad, religion is compelling. And it's not that just any strange belief will catch on. There is a limited catalog of possible religious beliefs that will last, because they can only be compelling in certain ways. The explanation of why that is so involves all the hypotheses in this book. I will show how certain mental disorders correlate with religion (mania, obsessive-compulsive disorder, schizophrenia, and temporal-lobe epilepsy all feature hyperreligiosity). Mental and brain disorders also might have influenced the evolution of religion in human culture. But the fact that all cultures have religion shows that one need not have a mental disorder to be religious. The famous Minnesota twins study found that 47 percent or more of religiosity is—brace yourself—*genetic*.[22] How you are raised only affects 11 percent of the variance of religiosity. Go ahead and read that again if you have to—I certainly did when I first came across it.

I will also spend a good bit of time discussing various religious beliefs and experiences, but I don't want to scare off readers who are religious. No matter what your religion is, there are many adherents to the literally thousands of *other* religions you *don't* believe in. I guarantee you that there will be some religions in the world that will sound bizarre to you, and you can legitimately wonder "how can people possibly believe that?" Even religious people wonder about the beliefs of *other* religious people.

*   *   *

Some people don't think we should study things like art and religion. The *frivolity critique* often comes from scientists who look down on studying the arts or religion. The *protectionism critique* objects to the project out of fears that its success will take jobs away from human artists. The *spoiler critique* holds that I will squeeze all the beauty out of art by understanding what makes it work. The *impossibility critique* is that the entire endeavor of understanding the workings of compelling experiences (particularly art) is impossible. I'll address these concerns here.

Religion and the arts are frivolous, unimportant things. Studying them is a waste of time. The creation and consumption of art comprises a major portion of our lives. Young Americans spend about seven and a half hours *every day* consuming artistic media. The creation of art is also important to us, as self-expression, as therapy, and as a way to attract mates. To get the stress-reduction benefits of five hours of post-work decompression, you only have to look at art for forty minutes,[23] and listening to music is the most popular means for relieving stress.

Anything we spend that much time attending to and that can affect us so greatly is worth some study.

Moving on to religion and other belief systems, ideas (both good and bad) spread like viruses throughout populations of people. Though belief systems involving the paranormal and quack medicine are easily dismissed by scholars, the belief in such things in the general populace is shockingly widespread. More than 40 percent of Americans believe in ghosts, devils, and spiritual healing.[24] Religion is found in every human culture, a feature that is shared with only a few other phenomena: language, tool making, and art (including music). Religious ideas are a major factor in the beliefs and behavior of peoples around the world.

Technology will take jobs from artists. Technological progress has put machines in the roles that used to be taken by laborers. Typically, this brings people out of physical labor and creates new white-collar jobs. We are already at the point where computers are starting to take over intellectual jobs as well. The original meaning of *computer* was a person who did computations by hand. This trend will surely continue, and people will be working in higher-level jobs that manage machines working at the lower levels.

I predict that as computers gain expertise in creative areas, particular jobs will start disappearing, just as cel animation (cartoons made by drawing rather than using computer graphics) is disappearing now. That is not to say that there will be no human-made art.

People will make it for the intrinsic pleasure of it, much like poetry is made today.

Will computer programs of the future be able to do everything the people of today can do, and better? I think so, but I also think that there will be a computer-machine synthesis that will render the human-computer distinction irrelevant. It also might be that by the time computers are better at everything than we are, there will be no need to work at all. I should note that there are smarter people than me on both sides of this debate as well. Time will tell.

If computer programs can create worthwhile works of art cheaply and quickly, art consumers (that is, everybody) will benefit enormously. Perhaps films can be customized to be exactly what I'd most want to see given the day I've had and feature themes reflective of ideas I've been wrestling with lately.

You're a spoiler: you'll squeeze all of the beauty out of art by understanding what makes it work. Poet John Keats lamented that Isaac Newton's theories had "unwoven the rainbow" and ruined the beauty of color and light through his investigations. Keats died in 1821, but in my (admittedly nonscientific) casual observation of arts and advertising, artists and designers are still using color with some effectiveness. This understanding of how something works offers its own kind of beauty. For example, people like landscape scenes that appear to be looking out from a partial shelter (e.g., tree branches). I think this is because the branches make us feel safe, which is a beautiful thought in itself, and enhances and deepens our appreciation of a given landscape painting or photograph. Is sex any less fun because we know it evolved for reproductive purposes? Are cupcakes any less delicious because we know we evolved to like sweets because sugar was nutritious and rare in our evolutionary environment? It sure doesn't feel like it. When I started writing this book I believed that explanation wouldn't change the feeling of appreciation.

However, one of the humbling things about putting your trust in science is how often you find yourself to be wrong: a study by business professor Sarah Moore showed that explaining why you like the taste of a cupcake diminished the love for the taste. She also found that explaining why an experience was horrible reduced the remembered horribleness.[25] Another study, supervised by psychologist Daniel Gilbert, showed that a complete understanding of a positive event reduces the felt pleasure, as well as its duration.[26] So, for example, if someone asks you why you liked a movie, simply trying to explain why will probably reduce your pleasure of having watched it. Understanding seems to reduce the emotional strength of an experience, which works in our favor for negative things (recovery from trauma), but can hinder our pleasure for positive things (like eating cupcakes).

I was wrong, and I cannot now write a knock-down rebuttal to the spoiler critique. The best I can say is that even though your pleasure might take a hit, I hope that the explanations themselves are sufficiently interesting that, given the gains and losses, it is worth it.

There is one kind of entertainment that is clearly hindered by understanding: magic tricks. The reason magicians don't explain how their tricks are done is because this knowledge ruins the effect of the trick. Your mind's search for understanding is part of why the trick is so enjoyable.

Compare this to how much *more* we appreciate some things when we find out how they work: the joy of discovery when we unravel a mystery, or when we learn that the pony fish illuminates its belly so as to be invisible to predators below.

Finally, there are those who think that an understanding of things like art and religion is *impossible*. We'll see.

1
HARDWIRING FOR SOCIALIZING

For all animals, living requires killing.

People, like ocelots, deer, and every other animal, must consume cells that are alive or were once living. In our ancestral environment, it was absolutely crucial that we were able to detect living things in our environments. What's that in front of us? Our minds have complex understandings of basic *kinds* of things. When we see something new, we quickly try to classify it as alive or not alive. If it's not a plant, it might be another person, friend or foe, or it might be an animal that we might hunt or that might hunt us.

In their natural environments, animals can be difficult to see. When our ancestors saw an ambiguous shape, which might have been a fallen log, or maybe a lion, the ones who erred on the side of caution tended to live to have kids who then inherited, genetically or culturally, the habit of erring on the side of caution. Nobody's going to die because they ran away from a log, thinking it was a lion, but somebody sure might die if they hang around a wolf, thinking it's a pile of leaves. Because it has been safer to assume that ambiguous things are animate, we often mistake inanimate things for animate ones.

Humans, like many of our primate relatives, live in complicated, hierarchical social environments, and have done so at least since the dawn of agriculture.[1] An individual's prosperity requires careful tending to a garden of allies, superiors, rivals, lovers, and enemies. Understanding the structure of one's social surroundings, and one's place in them, is crucial for happiness, reproduction, and even survival.

Understanding one's own personal relationships is social reasoning's simplest form. Even dogs know who their friends are. What is more complicated is knowing relations between two *other* people, such as the relationship between your friend and her spouse, let alone knowing who else knows about this relationship in your social circle. Think of all of the details a person needs to keep track of in his or her circle of family and friends.

Keeping track of all the social intricacies in a community requires a staggering amount of mental power. Those who could not understand and successfully manipulate their complex social ties were probably taken advantage of, outmaneuvered, and, ultimately, outbred by those that could. The people with social smarts lived.[2] It is likely that the importance of understanding social structure has, through evolution, resulted in a deep-seated, primal desire to pay attention to people and social relationships. Just as roasted meat smells good and safe places feel inviting, relationships between people are inherently, irresistibly *interesting*.

The thesis of this chapter is the *social compellingness theory*: our bias to perceive and be interested in people means that information about people and social relationships makes everything more compelling.

The theory has two parts. First, people tend to believe patterns have something to do with social meaning, intention, and agency. Sometimes this is called "agenticity," the "hypertrophy of social cognition," an "overactive theory of mind," the "hypersensitive/hyperactive agency detection device," or "anthropomorphism."[3] This is what happens when one sees a face on a Martian mountain or thinks that lightning is thrown by an angry god.

When you see something that has the potential to be very good or bad for you, such as a lion or another person, your mind goes through a series of processes. Your attention is drawn to what you see, and you think less about other things. You actively seek more information, through eye movements or by getting a better vantage point. You might have an emotional response, such as fear, anger, or desire. You might make plans, just in case.

But for human beings, this instinct goes beyond mere animism, that is, ascribing life to something that isn't alive, such as thinking that a log is a lion. Not only are we hypersensitive to detecting life and animate creatures, but we are hypersensitive to the presence of other people—human beings with minds.

Second, social explanations that we hear from other people are more believable. That is, explanations couched in terms of people and their interactions just feel right. When we hear about characters with desires, it resonates with something deep within us. For example, in an economic downturn, an explanation that places blame on a few people can be more believable than an explanation involving chaotic market fluctuations, because it is natural for us to see things happening as a result of human action. When soldiers torture, we find blaming a few bad eggs more satisfying than examining systemic problems that encourage certain behaviors.

According to social compellingness theory, we perceive people where there are no people and we prefer social explanations. So, how does social compellingness affect our love for the visual arts?

My friend Daniel Thompson is an artists' model. He poses so people can sketch him. For a while he brought his dog Phoebe along with him. He said the artists liked that; they would sometimes sketch Phoebe "as a palette cleanser" or might sketch the dog after they sketched him. But they never opted to sketch *only* his dog.

An art historian told me once that if you look across the arts of different ages and cultures, you find no common characteristics. I countered that nearly all of them feature representations of people, especially people's faces. If you walk through any art museum, the proportion of works featuring images of people is remarkable. Why should our species get such better coverage than dogs? Social compellingness theory predicts that we are very interested in people and the relationships between them. People like pictures of people and find them more memorable.[4] For the visual arts, this means lots and lots of pictures and sculptures of human beings. Ptarmigans do not pay commissions. It would be a very different world if they did.

Exactly how much do people dominate artistic depictions? I asked Carleton University art history student Rebecca Frerotte to record how many people were depicted in each individual work of art found in a large art history book, *Art Past, Art Present*.[5] Although I will use the word *paintings* as shorthand in this discussion, the book contained works of art in various media. Of the 420 paintings in the book, 333 had at least one person depicted. There were over *three times* as many paintings featuring people, at least for artworks famous enough to be included in an art history book.

We can predict more than this, however. We should see more paintings with low numbers of people in them than high numbers. One reason to expect this is because it's easier to paint one person than groups. This is also true for films, where casting is expensive. Second, the more people present, the harder it is to have a meaningful conversation. As you might have noticed at dinner parties, the maximum group size for a single face-to-face conversation is around five.[6] Assuming people are often attracted to decently sized conversational groups, we would predict that visual works of art would have between one and four people in them (four plus the viewer makes five).

The third reason we'd expect low numbers of people in paintings is because of our ability to do something called *subitization*. Subitization is the ability to rapidly know how many things are in front of us, without counting. Most people cannot reliably subitize more than four objects.[7] Things that are more easily processed are more pleasing to us, a notion I will discuss in detail in chapter 3. For paintings, people should prefer paintings with a subitizable number of people in them.

For all these reasons, I predicted that the greatest number of paintings will have one person in them and for the distribution to sharply taper off at four or five people. The data support this as well:

116 depicted a single person, 48 had two people, 26 had three, and it trailed off from there.

The next time you visit an art museum, you can marvel at how exquisitely adapted the artwork is for human consumption. Museums are full of people looking at people.

Group dancing, a kind of participatory art, is a nearly universal cultural phenomenon. People get together and dance together, ecstatically. It could be that group dancing (and for the military, synchronous drilling and marching) is an innate binding mechanism for groups.[8] In David Byrne's book *How Music Works,* he describes a similar effect caused by playing with a large band:

> The band became a more abstract entity, a community. And while individual band members might shine and take virtuosic turns, their identities became submerged within the group. It might seem paradoxical, but the more integral everyone was, the more everyone gave up some individuality and surrendered to the music. . . . As I experienced it, this was not just a musical transformation, but also a psychic one. The nature of the music helped, but partly it was the very size of the band that allowed me, even as lead singer, to lose myself and experience a kind of ecstatic release.[9]

Social compellingness theory predicts that narrative arts will also usually be about people and the relationships between them. We can have nonrepresentational art that does not depict people, but it is difficult to even describe what a narrative is without reference to characters. Our interests in fiction reflect our interests in real life. For example, just as we're interested in the making and breaking of friendships and romantic relationships between our real-life peers, a casual look at fiction reveals endless stories about friends and lovers. Cross-culturally, the most common narrative themes are love and conflicts between people.[10]

It makes sense that we care about the *real* people around us, but do we really care about people we know are fictitious? We certainly do. In the Western world, people spend an enormous amount of leisure time watching television and movies, reading books, and playing computer games, which mostly consist of stories about fictional characters. In fact, people will spend *more* time reading a text they believe to be fiction than the same text when they believe it to be nonfiction, suggesting that people are even more interested in fictional characters than stories about real people![11]

Not only are we interested in fiction, but we can be profoundly moved by it. People experience a range of real sensations from the shallow titillation of pornography to the soul-wrenching pathos of *Life Is Beautiful*.

When we experience a fictional narrative, the old brain and the new can give us different experiences. As described in the introduction, we are of two minds; we have "double knowledge." One part of our mind knows it's fiction, but much of our mind does not. Our emotional areas have very direct ties to these old brain areas, which explains why we can have such strong emotional reactions to books and movies. We take information in, and the critical evaluator (which determines whether or not what we're experiencing is real) and the emotional response run in parallel.[12] This is why your deliberate mind can mitigate, but not eliminate completely, the fear response to a horror film by telling yourself "it's only a movie."

This is also why we can get so attached to characters. In our minds, we feel we have become a part of the community we're experiencing in the narrative, even if they're vampires or wizards.[13] When we think that beloved television characters are going to leave a show, we anticipate negative reactions similar to what we'd expect to feel upon the dissolution of a social relationship.[14]

It turns out that our tendency to believe things we've heard to be true is old and a fundamental part of our minds. When we hear a statement, we believe it by default. When someone tells us they have had their colon removed, do we believe them? Unless we think of a reason to doubt people, we assume what they say is true.[15] Even when we know that we have been misled, that misinformation can still affect our reasoning, and we might act as though we still think it is true. Skepticism and the ability to critically evaluate and disbelieve probably came later in evolution. Disbelief is cognitively taxing, an afterthought that we often don't bother with.

We evolved to believe stories told to us, so at least a part of our minds treats heard stories as factual. We do this to gain knowledge about the world, and the emotional responses we have to many stories make them more meaningful to us. This is why we get weepy watching *The Notebook*.

If the origin of our affinity for stories comes from the drive to learn useful information, then it follows that the most compelling stories would be about subjects most relevant to the prosperity and reproduction of our ancestors. For example, stories about people like you should be more compelling to you, because they are presumably more relevant. Indeed, identification with the protagonist has been found to increase the degree of emotion experienced while reading stories, and, as mentioned in the introduction, news stories tend to be about evolutionarily relevant topics.

What kinds of things are we learning from stories?

Most narratives involve conflict between people, which resembles the environment we navigate in our real lives every day. Just as we strive to understand the minds of the real people around us, comprehending narratives requires simulating the complex social information they describe. Narratives can be seen as a primitive kind of virtual reality, making us forget our physical surroundings and feel as though we are transported into the world described by the novel, play, story, computer game, or film.

If you doubt the ubiquity of character interaction in narrative, I challenge you to make a compelling story about rocks and other nonliving things—without anthropomorphizing them. Fantasy stories imbue animals and inanimate objects with human goals, intentions, and personalities.[16] Their bodies might be different, but inside they are people, and as audiences we interpret them in this way. In one experiment by psychologist Ute Frith, squares and circles were presented to people, moving in certain ways. They were interpreted by the experimental participants as chasing each other, hiding, fighting, and so on.[17] Viewing these episodes activates the same part of the brain as when people are trying to understand the thoughts and motivations of others. That said, we do prefer actual social beings to simulated ones. (Earlier I mentioned that we prefer fictional stories to true ones—keep in mind, though, that even a character in a nonfiction story is simulated. The story might be *about* a flesh-and-blood person, but the reader is only experiencing a description in a book.) As wild and unrealistic as the worlds conjured by authors of fantasy, magical realism, and science fiction might be, nearly all describe characters that are *psychologically* realistic.[18]

Some authors, such as Stanisław Lem, deliberately try to buck this trend and introduce psychology that is completely alien to our own. Lem's fascinating novel *Solaris* depicts an alien entity that is completely incomprehensible to the reader. However, the novel also depicts human characters, who help us appreciate the alien's alienness through their own, equally baffled eyes.

Religion, like most fiction, imposes humanlike minds onto its characters, even if those characters are gods. Science, and especially science education, faces the daunting challenge of convincing people of the value of its stories, such as the origin of the universe, natural history, gene expression, and geological history. These stories are about true nonpersons, such as tectonic plates. Educators often need to communicate processes and sequences of events that superficially resemble stories but essentially have no characters. One way to deal with this problem is to anthropomorphize nonsentient things into real characters with beliefs and desires. For example, one might describe rain as *wanting* to *find* the *easiest* route down a mountain.

Even when communicating with each other, scientists use language rife with anthropomorphic metaphor. Stewart Guthrie's book *Faces in the Clouds* has a whole chapter full of examples. He even attributes part of the success of Charles Darwin's theory, in its time, to his anthropomorphic language—for example, his use of the word *selection*—which allowed Christians to read an intelligent designer into his theory of evolution.

Without characters, it might not even make sense to call these scientific explanations stories at all. But in the marketplace of ideas, they certainly compete with other explanations that *are* stories. I suspect that the lack of drama in scientific narrative explanation is one reason religious explanations can resonate with people so much more effectively. Worldwide, supernatural explanations overwhelmingly involve personified goddesses, gods, and spirits, all of whom have desires, personalities, and beliefs.

Even readers partial to science should not find it hard to imagine a person from a preindustrial culture being skeptical of a scientific explanation of disease. Imagine trying to convince someone that his or her sister is sick because of a germ that is so small you can't see it and that the sickness happened for no reason at all, rather than as punishment from a spirit for some wrongdoing or because of a curse cast by a rival.

Hunter-gatherer and tribal societies use witchcraft to explain every fatality that cannot be explained by an obvious physical cause. Strangely, even when they know about and believe in a physical cause, people *still* can believe in a supernatural explanation. That is, they believe in both at the same time.[19] In spite of the comforts that scientific explanations provide for these so-called random events, many feel that they just aren't enough. There is something about spiritual explanations that resonates deep within us and makes them difficult to resist. Part of what makes them so compelling is that they are essentially social explanations. We want *reasons* for things, not just causes. Our minds have evolved to think about intention and purpose, especially when reasoning about the origins of things. For example, one might know that germs cause disease but still believe that witchcraft explains why one caught a particular disease at a particular time.

It is interesting that people are willing to believe in invisible powers of witchcraft but might be skeptical of equally invisible germs when thinking about the causes of disease. It could be that we expect physical things to be visible, but we don't expect this of the supernatural.

We don't have to turn to societies that explicitly believe in witch-craft to find people who think about curses. A study led by psycholo-gists Emily Pronin and Daniel Wegner found that participants felt responsible for someone's headache when they were thinking bad thoughts about them. This effect was especially strong if the per-son with the headache was acting like a jerk.[20] It seems our minds naturally go to supernatural explanations. Religion is easy. Science is hard.

You might have noticed that many of these supernatural beliefs involve morality. Why might this be? One theory, proposed by psy-chologists Kurt Gray and Daniel Wegner, suggests that our minds look beyond the normal to explain why bad things happen to us. When someone does you wrong, you can hold that person causally and morally responsible. But when you are the victim of some ran-dom event, such as a lightning strike or getting sick, your overactive agency detector kicks in and posits the existence of either some su-pernatural agent (e.g., God) or some supernatural means for a nor-mal person to hurt you (e.g., a curse).[21]

Another theory, described eloquently by Jonathan Haidt in *The Righteous Mind,* and by Jesse Bering in *The Belief Instinct,* is that the tendency to be religious evolved to help group members treat each other better. Just as you're less likely to act badly if people are watching you, you're less likely to act badly if some god or spirit is watching you. This keeps you from acting selfishly. The group bene-fits, and the religious groups, over time, outcompete the nonreligious groups. This theory is contentious.

The techno-thriller writer Michael Crichton was routinely criti-cized for having shallow, poorly fleshed out characters. Indeed, his characters are often little more than interchangeable placeholders, distinguished by their abilities and roles in the plot rather than their personalities. Science fiction, fairy tales, and fantasy genres tend to have less developed characters. This presents us with a puzzle. If science fiction and fantasy have no deep psychological interest, then why do these genres enjoy such enduring popularity? Why would anyone watch *Star Wars* when they could watch *Do the Right Thing?*

What makes these stories compelling to the people who love them are the worlds they create. Often these stories occur in other worlds (either different versions of our world, or other worlds entirely) that work according to their own sets of rules. It's important to science fiction and fantasy fans that these stories are true to the worlds created in them—that the events follow the rules of the world the story has established. With plenty of exceptions, in other literary forms the stories are more "character-driven," that is, the motivations are more deeply psychological, and place greater demands on the audience in terms of understanding characters' minds.

Although they might not have deep characterization, many plot-driven stories involve multiple characters, deception, and hidden agendas: think of mysteries, crime and espionage stories, and thrillers. Comprehending these narratives requires keeping track of who knows what, who believes what, who's lying, who is suspicious, etc.

To help understand the draw of fantasy and science fiction, we need to consider autism-spectrum disorders. These conditions are characterized, in part, by problems with social interaction, as well as an affinity for simpler, rule-based systems such as chess, music, and computer programming. Interestingly, several symptoms of autism-spectrum disorders are similar to characteristics associated with the Western popular culture's nerd stereotype. Studies using the autism spectrum quotient (AQ) show that occupations commonly linked to the nerd stereotype are associated with elevated AQ. The highest average AQ is seen among computer professionals, mathematicians, and scientists.

I conjecture that high-AQ individuals will have a relatively higher preference for stories that focus less on character and more on plot and world building, as such stories pose fewer demands on understanding people. Strong preference for character-driven drama should be more appropriate for low-AQ individuals.[22]

Autism can be viewed as a problem of being *extremely* male. Most autistics are men. Consistent with this, women, presumably because they tend to be more interested in people than in physical systems, buy 80 percent of fiction.[23] Perhaps this is because men and women are socialized differently. However, a study by Simon Baron-Cohen found that even infants show sex differences in what they like to look at—male infants prefer looking at *things* more than female infants do and females prefer to look at animations of *people's faces* more than males do.[24] Recent studies suggest that not looking at faces might be an early indicator of autism in babies.

We've seen that narratives tend to feature people and I've suggested that lower AQ individuals will prefer deep characterization. We can also take lessons from anthropology to make sense of the number of characters in a narrative. At some point there are too many characters to keep track of.

In fictionalized stories based on actual events, one of the changes that is most often made is that a character in the story will be a composite of several actual people. Interestingly, true stories often have too many characters to be compelling!

Theatrical improvisation, or improv, is the collaborative creation of theater, made up on the fly, without a script. An improvised scene is almost always a narrative, featuring interacting characters. I was a regular performer of theatrical improv for almost twenty years. In a training course given by Joseph Limbaugh, I was taught many patterns that make for more entertaining, compelling scenes. One of particular interest is the concept of a "status challenge," which is when a character of lower status gains status over someone of higher status during the course of a story.[25] Status challenges are arguably the most important events that can occur in social hierarchies, as they involve a potential change in a society's power structure. Anecdotally, in my own experience as an improviser I have found that status challenges are the Midas touch for an improvised scene. We love to see apprentices outdo their masters or underlings getting promoted above their bosses.

Most narrative artworks are not interactive, but there are notable exceptions. Narrative computer games are interactive in the sense that the audience of the game, the player, plays the role of a particular character in the story—usually the protagonist. Similarly, players of role-playing games, both tabletop and live-action, play

characters with even more freedom. Doing so allows people to virtu-
ally experience stories as characters and explore the ramifications of
their choices. Although these games are not generally accepted yet as
art, I predict that computer games in particular will come to dominate other arts in the future.[26]

Financially, *computer games are already the dominant art
form.* People love watching movies, but Americans spend even more
time playing computer games.[27] The creation of many computer
games involves enormous teams of artists, including set designers,
character designers, story writers, voice actors, etc. They require
a production on the scale of opera in terms of complexity. People
start to get whiny if a film is longer than three hours. In contrast, if
a game takes fewer than forty hours to complete, it is criticized for
being short!

But even given all of this, today's computer games are primitive
versions of what they will become. It might seem hard to imagine
everyone playing computer games, and here is where science fiction
helps us with scenarios. I am particularly inspired by Neal Stephen-
son's vision of the future of interactive narrative as described in *The
Diamond Age, or, A Young Lady's Illustrated Primer,* in which interactive stories, called *ractives,* are enjoyed by individuals while the
supporting characters are graphic avatars remotely controlled by actors. In the book, a group tries to put on a traditional theater piece
and they have to keep asking the audience to stop coming up on
stage to interact with the characters.

What gives interactive fiction the potential to be more compelling than static fiction can be traced back to why fiction is interesting
in the first place. According to compellingness foundations theory,
fiction is riveting because it serves as a substitute for actual experience. We love to experience new things, because we learn from them.
But actual experience is more primal, and, in general, more compelling than vicarious experience. Interactive fiction returns the feeling
of actual experience to narrative in a way that static fiction will find
difficult to achieve, in the same way that a story about riding a horse
is less thrilling than actually riding one. While parts of our brains

believe the fiction is real, others don't, making the substitute not quite as potent as the real thing.

Simulations, games, and "choose your own adventure" books offer different paths through a world, where a book or film offers just one. One of the powerful aspects of these media is that one can return to them and see what would have happened if other choices had been made. Computer-game scholar Ian Bogost calls this "procedural rhetoric."[28]

Tabletop role-playing games (RPGs, such as *Dungeons & Dragons*) are plot-driven, interactive narratives that occupy the interesting space between literary fiction and theatrical improvisation. The typical RPG is played by two or more people sitting together at a table. One of the players is the game master and the rest of the players control fictional "player characters." Each player describes to the group the actions of his or her character and the game master controls what the characters experience as a result. This includes the natural elements of the world and all of the characters *not* controlled by players (nonplayer characters, or NPCs). Unlike many games, winning and losing are, for the most part, irrelevant concepts. The game master prepares an "adventure" and the entertainment of the game session involves playing out the story. Dice rolling is used to resolve conflicts such as combat, skill use, etc. For example, a game master might inform a player that there is a wall in front of his or her character. The player might announce that his or her character will try to climb the wall, and the rules of the game usually dictate what kinds of dice need to be rolled to determine the success of the attempt. Computer RPGs and most interactive fiction are like tabletop RPGs, except that the game master is a computer program. For choose your own adventure–style books, the game master is the book itself.

Although tabletop RPGs are compelling and popular, they are, relative to other narrative forms, severely understudied. Whether or not you consider role-playing games a form of art, the immersion people feel when playing these games is distinctly artlike. Although art scholars do research on interactive art installations, they have not touched the subject of RPGs.

In one aspect, RPGs have the potential, like the second-person narrative voice in fiction, to be even more immersive than other narrative arts: the events of the narrative might seem even more important if they are described as happening *to the reader.*

Other narrative voices require the reader to imagine himself or herself in the place of the characters, but second person does it for you. Some literary scholars claim that this is not so, that the second-person voice is off-putting, drawing attention to the words rather than their content. If they are right, it might be because the second-person voice is so rare as to draw attention to itself. Only by testing people who have habituated to it could we determine whether or not it is more or less immersive, independent of novelty. I know of no empirical studies of literature that have investigated the relative immersion of different narrative voices. However, we can get a clue of what we would expect based on a study that was done in another field.

When people *imagine* themselves doing something, they report different experiences when imagining from a first-person versus a third-person point of view. In your mind's eye, imagining opening a door from the first-person perspective would appear much as it would if you were actually doing it—your arm would extend from the lower part of the (imagined) visual field. Imagining it from a third-person point of view, in contrast, would appear as if you were watching a video that someone had taken of you. In a study by psychologist Lisa Libby, when people were asked to imagine activities like this from a first-person point of view, they reported low-level, sensory characteristics of the experience, such as "I feel the knob on my hand." When imagining it from a third-person perspective, they reported more on the meaning of what's happening: "I'm trying to get out."[29]

Guilt is one, possibly one of many, feelings that role-playing games can generate more easily than noninteractive narrative media such as novels or film. In most other media, audiences merely watch bad decisions play out. Role-playing games, be they on the computer or around a table, force you to choose character action, allowing you to feel pride or guilt for what you have done. In the computer role-playing game *Grand Theft Auto 3,* you can choose to make the character you control mug a prostitute and get your money back. Philosopher Grant Tavinor reported feeling guilty for what he had his character do to a fictional prostitute.[30]

Just as tabletop RPGs, literature, and the imagination offer different narrative points of view, so it is for 3-D computer games. In a first-person game the player's character is not visible. The character's gun, for example, usually can be seen sticking out from the bottom of the screen (e.g., *Doom, Quake*). These games are often called "first-person shooters" because of their focus on guns. Aiming is more intuitive with a first-person point of view.

In other games the player can see his or her character on screen. The on-screen representation of the character is sometimes called the "avatar," particularly when the character is supposed to represent, in some way, the player (e.g., *Tomb Raider*). Indeed, people tend to project themselves into these online characters. When viewing images of their avatars, gamers use parts of the brain used when thinking about themselves. This does not happen when they view, for example, images of their friends.

*   *   *

Our desire to learn about other people is evident in the compellingness of gossip. Cross-culturally, 80 percent of casual conversations is about social topics, according to anthropologist and evolutionary psychologist Robin Dunbar.[31] Almost everyone gossips once in a while. Although gossip suffers from a bad reputation, we often benefit from it, as gossiping forms a social bond, is 70 to 90 percent correct, and seems to prevent selfish behavior.[32] But the *content* of gossip reveals another of its purposes. Some of the same things that make narratives compelling are equally predictive of what makes gossip compelling. The content of gossip is often secret information, shared with a trusted confidant. Often we don't want the subjects of the gossip to know they are being talked about.

Why do people love secrets so much? In social environments, having secret knowledge gives one an advantage over others. Knowledge of, for example, a secret adulterous affair can give one a good deal of power over those involved. Gossip knowledge itself can help one predict future changes in the social hierarchy. It turns out that people are terrible at keeping secrets. In one study by psychologist Anita Kelly, 60 percent of the people asked confessed to sharing the secrets of their best friends.[33] Even telling someone that a bit of information is "secret" might make him more likely to repeat it, because that flags it as being more valuable!

This explains why so many compelling stories involve infidelity and why the plots of Shakespeare, for example, are rife with secrets—people seeing things they shouldn't, deals made in private, spying. The audience finds all of this thrilling, because their brains evolved to eat up this kind of information.

If gossip is so valuable and interesting, why, then, are spreaders of gossip frowned upon in almost every society? While we often like to hear gossip, we find gossipers a bit less trustworthy. After all, our own secrets might not be safe with them. Also, to outwardly disapprove of gossip might signal to others that you are unlikely to gossip and are therefore more worthy of being trusted with gossip-worthy information.[34]

*   *   *

The social compellingness theory holds that our interest in people comes from a need to understand the social world we live in. One prediction the theory makes is that we should be most interested in information regarding people who matter most in our lives, such as lovers, rivals, enemies, leaders, and those we trade resources with.

If it's important to know who is below us and who is above us on the social ladder, then we should be most interested in information that pertains to our own potential status changes. A study by psychologist Francis McAndrew found that students were not very interested in awards their professors got, but the same information about their friends and romantic partners was very interesting and more likely to be shared.[35] We are very interested in negative news about those who are perceived to be higher than we are, the great ones who might fall, and positive news about those below, the up-starts who might threaten our status.

How we react to people depends on how far we perceive our-selves to be above or below them on the social ladder. I call it the *relative social status hypothesis.*

For those *much lower* in the social hierarchy, we tend to respond with care. They are so far below us that they pose no threat to our place on the social ladder. We are their superiors and, when we think of them at all, we feel responsible for them, perhaps because of a subconscious interest in gathering allies.

For others *slightly lower* than us, we tend to respond with some care, tempered with the careful observation one gives to rivals. We are their superiors, in our minds anyway, but they remain rivals. Any sign of their ascension is a threat to our status.

For others *roughly equal* to us, the relationship is unstable and competitive. The others could be friends, possible allies, or possible rivals. Good or bad things about these people will always be inter-esting, because any change in status for them means a change in our own relative status. The relative social status hypothesis predicts that we would be *most* interested in gossip about people roughly equal to us in social status.

For others of *slightly higher* status, we tend to respond with ag-gression. They are higher than we are in the social hierarchy and we might be about to overtake them. We will delight in their downfall, as it would mean a gain in status for ourselves. We feel aggressive because by taking action we might be able to facilitate this downfall.

We treat others *much higher* than we are as role models or he-roes, and we have a tendency to ingratiate ourselves to them. They are so high above us on the social ladder that we see no hope of overtaking them. Given that they likely see us as someone to protect, they are not a threat to us either. The best strategy is to befriend and respect such people.

In general, people are more competitive with people of their own sex and age, so these predicted reactions should be strongest in these situations.

Of course, these predictions are for general reactions. In specific contexts other factors are at play. If someone is your good friend, you might not be aggressive to her just because she's slightly higher than you on the social ladder. But I'd predict that you'd be *more* aggressive than you would be if she were slightly lower than you.

Gossip has another societal function, however, and that is keeping people from acting badly. Most of our evolutionary history was spent in groups of about 150 people, and good reputations were very important because they were necessary for cooperation and alliances. Doing something that hurts the group tends to hurt one's reputation in the group. For a social species with language, if *anyone* sees that you are doing something wrong, the whole tribe is likely to catch wind of it. Psychologist Jesse Bering believes that this is one of the reasons the idea of a god culturally evolved.[36] We need to inhibit our selfish desires in order to live productively in a society. Being observed by people helps. But our ancestors needed additional control. The idea of an all-knowing god who watches us all the time gives us an additional inhibitory motivation. Without this, people would be unable to exert the control they need to be consistently kind—to not act entirely selfishly.

Without language, the only way someone would know if you did something wrong was if they directly perceived you doing it. With language, one bad act can tarnish someone's reputation in the minds of everyone who hears about it—for years.

People tend to think that supernatural agents have strategic, gossip-worthy information. According to social scientist Diego Gambetta, this information focuses on other people, is secret, and centers on status, sex, and resources.[37] Even when theological doctrine holds that a god is all knowing, people are still more likely to find statements containing socially strategic information (e.g., "God knows you cheated") more natural than sentences such as "God knows how often your dog eats," according to research by Benjamin Purzycki.[38]

Gossip has been a way of communicating about other people for tens of thousands of years. Now technology allows us to gossip more efficiently. I'm talking, of course, about the news.

Many people aspire to be knowledgeable about the state of the world. For most people this requires, at minimum, paying attention to news sources. There is an assumption that consuming various news sources will provide us with the information we should know—the important information.

Most news organizations, however, are not rewarded for providing people with the most important information. Profit-driven news organizations need people to watch the news, ultimately so that they can get subscription or advertising income. Even public news sources will use a large audience to justify public funding. All these organizations are rewarded for larger audience sizes.

This would not be a problem if we, the news media audience, were most interested in exactly those things that are most important. Unfortunately, the overlap between the most interesting things and the most important things is imperfect. Most news sources seem to value interest much more than importance. To survive in the competitive marketplace of ideas, they cannot afford to act otherwise. Since cable television came on the scene, television news, for example, competes not only with other news programs, but with entertainment. How do TV journalists make news interesting enough to compete with situation comedies and high-quality dramas? By making the news as story worthy and dramatic as their sense of professional responsibility will allow. The nature of news stories is one of the clearest examples of the social compellingness theory at work.

Apropos to this chapter, news items are called "stories" for a reason—people find interesting those narratives that are populated with characters. Even the stock market is more affected by newspaper articles that have photos of people with them.[39] However, it's been shown that when it comes to hearing about real events, people are most sympathetic when hearing something sad about one person, and for every additional person, the sympathy goes down. When we hear about massive deaths, we paradoxically find it less tragic.[40] This was summed up in a quote misattributed to Stalin: "The death of one man is a tragedy. The death of a million is a statistic."

The next time you read a newspaper you can marvel at how it is tailored to speak to our deep desires to know what's happening to, and in terms of, people. We like to hear about very tangible wins and losses, but we also care about symbolic ones—such as wins in sporting events.

The social compellingness theory suggests that sports are riveting because of the conflict between people that these competitions represent. Watching these competitions play out is compelling for the same reason conflict is compelling in fiction: we want to know who will win.

It is safe to say that we evolved to pay attention to competitions over actual resources—competing over a mate, over food, or over social control. But why would we care about artificial competitions, like basketball or chess? When we watch sports such as boxing, we know that the boxers' intent is not to kill. We are even more certain of this for table tennis. However, I don't think that we find sports competitions compelling merely because they remind us of real conflict.

Nature is full of symbolic competitions. Male bower birds, for example, don't fight over females, they try to impress them with the most elaborate homes. Many animals do a lot more screaming and throwing around of brush than actual fighting. The animals use signals representing strength or the ability to gather resources. These behaviors are the animal equivalents to sports and other competitive games.

In humans, different cultures have different symbolic competitions for resources. Americans compete for leadership by accumulating votes. Success in competition, including sports, is indicative of some more meaningful superiority, beyond the skill required to be good at the competition itself. This quality might be strength, or intelligence, social power, or humor. Ultimately, competitions like sports are signaling culturally or evolutionarily instilled values.

Why do we love to watch or hear about people doing things that are difficult, or doing things extraordinarily well? Parts of our minds think that every person we see belongs to our community, be they

on television, on an Internet video, or in a circus. We find it compelling to watch them because they could help us using their unusual
capabilities, or because sharing that information might prove useful
in bonding or negotiation.

Our love for virtuosity explains why we have an interest in people breaking records in sports and other games, even if the record
broken is the athlete's own. There are noncompetitive sports, but it
is worth noting that they are rare and less popular with spectators.
Conflict is compelling.

But what about music? The social compellingness theory predicts that the music that *does* have something to do with people will
be more compelling. For example, vocal music features the human
voice, and as such we'd predict that people would like it more than
instrumental music.

There is a debate over the "content" of instrumental music, but
music with lyrics overwhelmingly involves human themes—particularly sexual and romantic ones. A study by psychologists Dawn
Hobbs and Gordon Gallup found that 92 percent of the most popular songs (from 2009) contained "reproductive" messages, with an
average of 10.49 per song. Bestselling tracks contained more. These
embedded messages have been found in songs dating back 400
years![41]

*   *   *

Our affinity for thinking about people makes us more likely to generate and believe social explanations. For example, people don't just
want to know the *cause* of sickness, they want to know the *reason*—
the moral or social explanation. If we hunger for social explanations,
conspiracy theories are the hot fudge sundaes of the explanatory
world. I have noticed that simply explaining a motive for wrongdoing is enough to convince some people that the parties in question
are guilty (for example, in the 1960s, Americans wanted to scare the
Soviets, which conspiracists saw as a compelling enough motive for
them to fake a moon landing). Why are conspiracy theories so interesting, and so compelling? Conspiracies have two characteristics

that people just love—relationships between people and the secrets they hold.

Our predilection for social explanations makes us susceptible to believing things that are in conflict with scientists' views. Nearly all theories commonly referred to as "pseudoscience" have something to do with people, be they us or ancient peoples. Astrology, for example, suggests that the positions of the stars affect the personality and fate of people according to their birth date. Why should the layout of the stars have *anything* to do with human personalities? Quite simply, we like theories that center on us, that make human beings relevant. As my screenwriter friend Jennifer Mulligan says, "Astrology is a way to dramatize the sky." The popularity of horoscopes is an excellent example of our species centrism. It is quite rare to find a ritual regarding a supernatural agent that does not involve some kind of expected interaction—for example, to hurt someone, to bring rain, to remove sickness, etc.[42] Our ideas about normal human reciprocity and exchanges of goods and services extend to gods and spirits.

One exception to the trend of supernatural beliefs having to do with human affairs concerns the "high gods" that tend to appear in societies without writing or a complex social hierarchy. Such societies will often have a god or set of gods that supposedly created the universe, but those gods are often not objects of worship and are said to have little interest in human affairs. However, even though these high gods might not have much to do with human affairs in particular, they are, like almost all conceptions of gods, humanlike, or at least animal-like.

These patterns suggest that people would find explanations couched in terms of characters with beliefs and desires to be more compelling, and more believable, than explanations that have nothing to do with humanlike desires—let's call them "mechanistic" explanations. One aspect of this is a preference for agency—that characters do things on purpose. News reporters are more likely to talk of a stock price change as agentive when it moves in a steady direction. Business researcher Michael Morris found that agent metaphors make viewers assume stock trends will continue.[43] Further evidence for this idea has been found with a study by psychologist Deborah Kelemen. She found that children tend to prefer teleological (goal-focused) and functional explanations (explanations involving reasons) over non-functional ones, regardless of how religious their parents are.[44] For example, children might describe the sun as shining because it needs to heat up the day. Animate entities seem more conceptually accessible and are therefore easier to understand, and thus are more likely to be believed. I will explore the relationship between ease of mental processing and believability in detail in chapter 3.

Cognitive scientist Jesse Bering has noted that creationists will probably always outnumber believers in evolution because people have evolved to think in terms of design and purpose, especially when we see something that appears to be ordered. Ironically, we evolved in a way that prevents us from believing in evolution! [45]

In my laboratory, we tested the general idea that social explanations are preferred. We made up physical phenomena about made up concepts and offered two differing explanation types: a "mechanical" one that explained the phenomenon in terms of forces, and an "anthropomorphic" one that explained it in terms of the beliefs and desires of agents. For example, an anthropomorphic explanation might read "the dax comes to the jey because they want to be together." We predicted that the anthropomorphic explanations would be more believable than the mechanistic ones. To maintain scientific integrity I have to say that we failed to support this hypothesis. It appears that undergraduate psychology students, anyway, prefer mechanistic explanations to anthropomorphic ones.[46] When I presented this work at a scientific conference of experts on explanatory reasoning, they were just as baffled with the results as I was.

*   *   *

Our preference for social thinking makes explanations that make people the most important thing in the universe very attractive. Modern cosmology shows that we are not, but maybe, some believe, aliens are—this is the extraterrestrial hypothesis. There is a disturbingly widespread belief that intelligent extraterrestrials abduct people to perform medical-like examinations on them.[47] Social groups of so-called abductees have shared their stories and developed a subculture with its own mythos, including different alien types with different roles. What we now think of as the prototypical alien (naked, large head, large slanted eyes, small mouth, small or missing nose) is considered by the abductee subculture to be a "grey," and believers discuss the greys' nature, the greys' motives, and probably the greys' anatomy.

The most striking aspect of these stories is how much the greys look and act like humans. They are bipedal, on average about the size of a human woman, and bilaterally symmetric. They have recognizable eyes, heads, arms, legs, hands, skin, and (sometimes) mouths. Although they sometimes have distorted versions of human body parts, those parts are recognizable and in the same places on the body.

Psychologist Frederick Malmstrom has suggested that the face of the grey looks the way a female face looks to a newborn baby. [48] If this is true, perhaps we find the face compelling in part as a kind of primitive nostalgia for the face of our mothers. Newborn babies recognize faces using a very primitive part of the old brain—the hippocampus. It appears to be hardwired. As babies develop, they use different parts of their brains to recognize faces. One aspect of this old face-recognition system is that it does not use the presence of hair or ears to detect a face.

Note that greys are depicted as having no hair and no visible ears. Newborn babies have vision that is coarser than that of adults, which results in the loss of detail required to see the nose and mouth clearly—they vanish or become slits. What a baby sees can be imitated by blurring an image, which in adults can actually improve facial recognition.[49] Further, newborns see the world in shades of grey, which suggests a reason why we'd find the idea of grey-colored aliens compelling.[50]

Using software, an image of a woman's face was manipulated so that it would resemble the way a newborn would see it. The whole picture went into grayscale, the eyes became narrow and diagonally slanted, the nose and mouth became less prominent, and the ears and hair disappeared. It looked strikingly like a grey alien.

Not only do aliens look like humans, they *act* like them. According to the abduction narrative, their goals are recognizable, and in fact similar to some human goals. The stories told often involve sex and violence. They perform bizarre sexual experiments on people involving pregnancy, reproductive organs, intercourse, and so on. Journalist Kaja Perina reports that 60 percent of female and 50 percent of male alien abductees claim to have been examined by the aliens while lying naked on a table.[51] Stories having anything to do with mating, such as sex and romance, are more compelling.

Although the extraterrestrial hypothesis is not normally thought of as a conspiracy theory, a bit of conspiratorial thinking is required for it to make sense. After all, there is no incontrovertible evidence that aliens have visited us. One might think that with the thousands of people being abducted, and with all of the implants the aliens have supposedly been putting into people, *some* piece of extraterrestrial material would eventually show up and settle the matter. But it hasn't, so skeptics don't believe. But what's a believer to do?

Enter conspiracy. First, the aliens themselves supposedly don't want us to know about them. Fine. So believers in alien abduction introduce evidence of the cover-up into the story. The abduction narrative includes things like memory wipes and implants that look just like normal material that we could find on earth. The alien-abduction theorists think that the aliens are advanced and cover their tracks, but they are apparently sloppy enough to leak enough clues to convince a sizable human community of believers.

Further, the extraterrestrial hypothesis holds that human governments are covering up the evidence that already exists. Special agents are said to exist who go around hiding it all. Often UFO sightings are later claimed by the government to have been flights of experimental aircraft that were kept secret. Believers say that the government wants you to believe in their secret aircraft so people won't know the truth about aliens. Skeptics say that the government is happy that people believe in aliens so they won't pay much attention to their experimental aircraft.

What is the government's motive for hiding the truth from the public? The suggested answer is that the government believes people would panic if they knew the truth. Perhaps this made more sense to the mind of someone in the 1950s, but today, the idea of widespread panic seems very unlikely. Think of all the people that would have had to be involved, over the course of 60 or so years, to keep alien contact a secret. It strikes me as very unlikely that *not a single person involved* would blow the whistle on this thing in 60 years. Conspiracy theorists of all stripes generally overestimate people's ability to keep their mouths shut.

Believers say that the sheer number of reports means there must be some truth to the alien theory, but the nature of this narrative has much more to say about human psychology than it does about extraterrestrial life.

* * *

Some have likened belief in alien abduction to religion, and indeed religions the world over contain beliefs about supernatural beings with desires and personalities. The most obvious personified forces in supernatural beliefs are gods.

Let's look at other aspects of the occult and compare them to scientific theories. Some of the beliefs popular with new-age religions include: crystals can channel life force; energy is passed between people; people have auras that are different according to mood and personality; the position of heavenly bodies, relative to earth, have an effect on your personality and future; souls are reincarnated after death. All these theories have one thing in common: they all involve people. Why are there no supernatural beliefs about why mitochondrial DNA is not passed down in the same way as the rest of our genetic code? About why steel is stronger than wood?

Now, certain religions do have beliefs about nonhuman things, such as the cycles of the moon, harvests, etc. However, it is interesting that religions that involve rituals based on seasons, or the calendar in general, are mostly absent in societies without farming. Once a society becomes agricultural, suddenly calendrical rituals pop up.

Why? Because supernatural explanations are generated and maintained, in general, when they are relevant to human lives.

Religious explanations tend to be based on the divine will of gods or spirits, which are, basically, people, with their own opinions and motivations. Social compellingness theory predicts that religious beliefs that reject personhood in the supernatural in favor of nonagentive entities (such as a nonanthropomorphic energy) will have more trouble surviving and will evolve (culturally) into more successful, anthropomorphic versions, if they survive at all. One bit of "evidence" for this is that in all (or nearly all) religions, people believe in supernatural agents. Of course, if these agents are a part of religion's definition, it's not saying much to claim that all religions have them.

Our ability to reason about what other people are thinking is known in cognitive science as "theory of mind." Cognitive scientist Jesse Bering suggests that much of our religious belief comes from an overactive theory of mind applying itself to places where there are no minds at all,[52] which is also the first part of social compellingness theory. Although some people might prefer to talk of God as being some nonpersonified force, when people pray, the same parts of their brain are active as when they are interacting with other people.[53] At a perceptual level, we can mistake nonanimate things for animate things. We might see a garbage bag as a crouched person or see a face in the headlights and grill of a car, but it is rarely the other way around. In addition to the perception of humanlike forms, we also have a natural tendency to infer that other things have *minds* like ours.

Anthropologist Wendy James describes a cult in Sudan that believed that ebony trees could hear people's conversations and would sometimes reveal what they heard.[54] What is also notable, however, is that the cult focuses on the trees' observations of *people,* as opposed to the myriad other things that could be observed, such as changing cloud patterns. Also, this religious idea would not be as compelling if the information the trees gathered was never revealed. Of course the conversations that matter are the socially strategic gossip-worthy ones. Religions focus on supernatural agents' knowledge about people, knowledge that can in turn affect human affairs, be it through divination, wrath, curses, or something else. Not all gods are believed to have moral (socially strategic) knowledge, but the ones that do are offered sacrifices. According to research by anthropologists Stephen Sanderson and Wesley Roberts, gods (in many religions) that do not have or share strategic knowledge do not have rituals dedicated to them.[55]

One thing that is particularly notable is that even if you are not in this Sudanese cult, and even if you don't believe in anything supernatural, you probably have a pretty good idea of what it's like for something, even a tree, to hear a conversation, based only on my very short description. This is because we all have similar representations of minds and plants, and when I combine them, or you combine them, or the members of this Sudanese cult combine them, we get similar results—similar concepts in our minds. Ideas of supernatural agents can be communicated very efficiently because every person more or less correctly reconstructs the idea in his or her own head.[56] These entities are made of relatively simple conceptual building blocks that we already have.

Some religions claim that God is not a humanlike entity, but some kind of force. This conception of a god is not common in religion, and seems to be a function of intellectualizing by religious authorities. In these religions, the belief that God is not humanlike might be "theologically correct," in that it is in agreement with what religious authorities say, but it does not accurately reflect the beliefs of laypeople. It turns out that laypeople might express theologically correct ideas when asked about religious theory generally, but when you ask them to interpret specific situations, people show their true theologically incorrect colors. Catch them with their guard down, so to speak, and God becomes much more anthropomorphic. Psychologist Justin Barrett reports in his survey of cognitive studies of religion that people might, when asked, *claim* that God can listen to many things at once, but misremember stories as saying that God could not hear something because there was a loud noise at the time.[57]

Children are particularly susceptible to seeing agency in inanimate objects. Elementary concepts such as agency are among the first to be acquired but are frequently overgeneralized to inanimate things. At four years of age, children seem to believe that *all* motion is intentional. But it seems that even many adults attribute agency to things that are *apparently* self-propelled: certainly animals, but also wind and astronomical objects like stars, planets, the sun, and the moon, which seem to move through the sky without anything pushing them. These beliefs turn up in religions.

I predict that people more disposed to thinking socially will be more likely to perceive animacy where there isn't any. If we look at extremes of social thinking, we see some evidence in support of this view. One extreme of social thinking is autism, which, as noted earlier, is characterized by a decreased ability to think socially. I don't think it's a coincidence that people with autism-spectrum disorders also tend to be nonreligious, according to Catherine Caldwell-Harris.[58] The impaired ability of people with autism to use "theory of mind" to understand the mental states of others might also explain why they tend not to believe in gods. Some have said that psychosis and autism are at opposite sides of a spectrum (this theory is not currently accepted by mainstream psychiatry). People with psychosis often see personal meaning in lots of random events.

Sociologist Fred Previc has shown that women worldwide tend to be more religious than men (there are exceptions for some regions and religions) and in general have more paranormal experiences.[59] This is possibly a side effect of their heightened abilities to reason socially. To test this, we would need to find a correlation between an individual's social-thinking tendencies and religiosity. There is preliminary support for this hypothesis in a study that showed a significant correlation between self-reported religiosity and emotional intelligence in Christians, and that autism-spectrum measures correlate with reduced belief in God.[60] Of course, the fact that religion seems to have a global hold on humanity, even in the parts of the world where men are in power, indicates that religion is far from being a strictly female phenomenon.

In this chapter I've examined a few related cognitive processes that are implicated in our beliefs about supernatural agents. I've described them under the umbrella of social compellingness theory, but they likely constitute a variety of functions. For example, there is a specialized place in the brain that handles the detection of faces. This *face-detection* process explains why we see faces in clouds or toast. But there is a different process being used when thunder is attributed to a god, because there is no experience of a face. *Animacy detection* is used in the scientific literature to mean the perception of something that moves on its own accord—unlike rustling leaves, for instance. This mental function evolved to detect predators and prey. If we hear a rustling in the leaves, it behooves us to know if it's something dangerous or just the wind. Steven Mithen and Walter Burkert suggest that our intuitive processes regarding predation inform our beliefs in supernatural agents.[61] Although this theory is speculative, there are intriguing bits of support. There appear to be lots of hunting metaphors in religious stories; experiences with supernatural beings are often frightening, and further they often involve being able to see but not hear, or hear but not see the agent—situations that are particularly salient and frightening when facing a predator. One interesting way to study this would be to survey gods who are pictured as animals and rate whether they are predators (carnivores and omnivores, perhaps) or prey (herbivores).

Agency detection is perceiving that something has a will and can take action—that something has a mind. Animacy and agency are often used to mean the same thing, though in certain cases, such as the closing of a Venus flytrap, something can be animate but not be an agent. *Theory of mind* is our ability to reason about minds. Theory of mind can be said to use *folk psychology,* though this term is sometimes reserved for our explicit ideas of how minds work, rather than how we reason about them intuitively. Agency detection is a part of theory of mind.

But keep in mind that just because scientists use different terms, it does not necessarily mean that the terms refer to different functions in the mind. It could be that many terms are just different uses of the same mental machinery. Psychologist Adam Waytz found that anthropomorphism activates the same brain area implicated with social cognition in general,[62] and psychologists Lasana Harris and Susan Fiske found that this area is less activated when thinking about certain groups of people as being less than human.[63]

In the section on gossip above, I described the relative social status hypothesis, and noted that we are most interested in gossip about those people who are similar to us in sex, age, and social standing. It turns out that this has religious ramifications as well. Some cultures have a belief in the "evil eye," a curse brought on by envy. However, the evil eye is not often thought to be cast by a poor person on a very rich person. The evil eye tends to be brought up in the context of social and economic equals when one of them is perceived to have gotten ahead in some way (given birth to a beautiful child, or received a sudden windfall of money). Pascal Boyer hypothesizes that this belief system piggybacks on our "cheater detection" system. If someone is pulling ahead, perhaps it is because they are magically stealing something from others.

When a religion gets very popular and starts to spread to different cultures, it often must somehow deal with the local god and spirit beliefs of the new converts. Often the spreading religion demotes these gods and spirits to lesser beings with respect to the bigger, more universal gods. Hinduism did this particularly effectively, incorporating the idea of local gods into the basic doctrine. Many heroes and even gods of the ancient Greek religion were appropriated as saints in Christianity. If the spreading religion fails to incorporate locally believed supernatural agents, many people will continue to believe in them anyway, outside of the scope of the spreading religion. The people in charge of religious doctrine would often like ordinary believers to be a bit more theologically correct, but it is a testament to the strength of our theory of mind, and possibly human nature, that these kinds of beliefs keep coming back.[64] People will continue to believe in gods and spirits that are local, and will even believe that the dead have interactions with their day-to-day lives.

Our instincts make us want to pay attention to people and social interactions. As shown in this chapter, this has two important effects. First, we overextend our social thinking into places where it is inappropriate, resulting in anthropomorphizing inanimate entities. Second, we find people fascinating and prefer them in our arts and explanations. An alien anthropologist looking at our arts and religions would have no trouble understanding humanity. Our natures are unambiguously inscribed on all of it.

2

WIZARD'S FIRST RULE

Hope and Fear's Anti–Sweet Spot

In Terry Goodkind's fantasy novel *Wizard's First Rule,* the wizard's first rule is that people have a tendency to believe things they hope or fear to be true.[1] The protagonists use this rule to understand and manipulate others in the novel. In real life, too, hope and fear are foundations of compellingness. Let's talk about fear first.

Drinking from glass containers is dangerous, because minute bits of glass naturally chip off from the jostling and bumping of the container. These sharp pieces get into your intestines and slowly, over time, tear the cells of the digestive tract.

If you don't know a lot about materials science, you just might have believed that. Even if you didn't believe it, you might feel the pull, the temptation, to take the warning seriously. But before I get into the reasons why, please know that I made all that up, and as far as I know glass containers are close to the safest things you can drink from. Cheers.

Fear makes us pay attention, and makes us believe things we otherwise might not. Fear isn't just feeling scared—it changes our whole mental state to prepare for danger. It shifts attention, heightens perception, affects goals, and primes certain actions, such as running or fighting. According to the idea of negativity bias, we pay more attention to, and are more likely to believe, negative information.

Why would we do this? As mentioned in the last chapter, one of the reasons language is extraordinarily useful is that we don't have to experience everything ourselves—we can hear someone else's account of it and form beliefs about what we heard. Epistemologists call information that is transferred in this way "testimonial belief." If you *disbelieve* what you hear, testimony won't help you very much. We all know that we shouldn't believe everything we hear, but it turns out that we do, unless we have some specific reason to doubt what we hear or read. This is sometimes called "the default rule."

Communication is especially useful for conveying information about danger. If you hear that there are venomous snakes on a certain mountain, you can just stay away. This bit of testimony just might save your life. Of course, soon after evolution gave human beings the ability to communicate, the ability to deceive people was greatly enhanced. Perhaps the person told you that there were snakes to keep you from some resource she discovered there. Nonetheless, it's not hard to imagine how the people who believed the reported danger survived and the ones who ignored warnings didn't.

We seem to have "better safe than sorry" built into our cognitive architecture. We have a hyperactive danger detector. Psychologist Maha Nasrallah found evidence of our sensitive danger detector in the fact that we are faster at detecting negative words (such as *war*) than positive ones. People have been shown to be more likely to believe generalizations if they are about dangerous things, according to a study by psychologist Andrei Cimpian.[2] In support of this evolutionary explanation for the fear bias, some things seem to be innately frightening. More accurately, it appears that we have a predisposition to learn to be frightened by certain stimuli more than others. Phobias, for example, can emerge after a single bad experience and can be very hard to cure. The most common phobias are of primal dangers: closed spaces, heights, snakes, and spiders. Strangely, the most dangerous things in modern life, such as cars and knives, are rarely the objects of phobias.

Snakes are a particularly interesting example, because some animals need *no exposure to snakes at all* to fear them. Even though there are no snakes in New Zealand, and there never were, pigeons fear rubber toys of them, presumably from evolutionary pressures from the places the pigeons' ancestors lived before the birds migrated to the islands.[3]

Psychologist Susan Mineka studied monkeys' reactions to snakes. In her lab, she raised monkeys that were never exposed to any outside monkeys. Her monkeys were not afraid of snakes. Indeed, they would reach out and try to play with them. She found, however, that it was very easy to teach them to be afraid of snakes. All that it took was showing them a film of another monkey reacting with fear to a snake. Interestingly, she could not teach them to be afraid of flowers, even when the flower replaced the snake in an otherwise identical film clip.[4]

What is going on here? It appears that there are certain "intuitive" concepts that are easily learned, even under less-than-ideal learning conditions. Nonintuitive concepts are difficult to acquire, even under ideal learning conditions. This appears to be an example of the controversial Baldwin effect, named after its creator, psychologist James Mark Baldwin. This is a theoretical evolutionary process in which what is passed on is not a particular psychological disposition but a mental state that can *learn* that disposition very easily. In this case, Mineka's monkeys were primed to be afraid of snakes.[5]

These examples show that fears can have evolutionary effects as well as effects on our beliefs. And what people believe, and why, can have serious real-world ramifications. Where people might otherwise think critically, their fear bias results in their accepting mere rumors as truth. For example, people will often believe hearsay regarding disease-causing agents.

Fear even affects our political affiliation. Conservatives are more sensitive to disgust and more interested in purity. Their eyes linger longer than those of liberals on repellent images, such as excrement and car wrecks, as shown in experimental research by political scientist John Hibbing.[6] In fact, a survey by psychologists Ian McGregor and Paul Nail showed that even making people more frightened in the laboratory makes them temporarily more conservative in their views, and on average people became a bit more conservative after 9/11.[7] Why? It could be that being fearful makes one risk averse and more likely to advocate tried-and-true solutions and courses of action.

*　*　*

The flip side of fear is hope. Although it's strange to think of it this way, one of the ultimate reasons we do anything is so that we will have beliefs that make us happy—for example, when you have a good job, you are happy because you believe it to be true. Usually, we come to have such beliefs and happiness by achieving our goals. Suppose we have a goal to walk outside. As we stroll through the forest, our perceptual systems indicate that we are indeed outside. This is rewarding because we had a goal to be outside that is now satisfied. It's not *being* outside that gives us pleasure, technically, but our *belief* that we are outside. People with delusions of grandeur get the happiness of believing that they are, perhaps, rich, when in reality they are not. Similarly, there are people who have objectively good lives but, for whatever reason, don't believe it, and don't enjoy the happiness normally associated with a good life.

Being productive gives us a rush of satisfaction. Drugs that give us a pleasurable high allow us to feel the rush without having to actually accomplish anything. Taking recreational drugs cheats the part of our minds that drives us to do important things.

People naturally like sweet tastes. Eating sweet things is pleasurable. But we can fool our taste buds with artificial sweeteners that contain no sugar at all. Just as an artificial sweetener fools our mind into thinking there's sugar without giving us the nutritional benefits of sugar, our minds can latch onto ideas we want to be true even when the evidence is poor.

Suppose a doctor informs someone that they are very ill. The person has an active desire to be well, and this desire will not rest until he can believe that he *is* well. The desire will be satisfied no matter how the belief comes about. Perhaps there's a medical cure that will allow the patient to get better, to reach his goal of health.

However, the process of actually getting cured might be painful, expensive, time consuming, or straight-up impossible, thwarting the person's goal to be well.

Now imagine a well-meaning friend (or a greedy charlatan) tells him that she believes the doctor is wrong. Perhaps she tells him that instead of undergoing this treatment, all he has to do to be completely cured is to bathe himself in the Nile River. Now the sick person must make a choice: to believe his doctor, who is telling him he needs a painful treatment, or his unqualified friend, who is telling him he can just swim in the Nile. Whom does he believe? Let's weigh the pros and cons for our hypothetical patient.

What are the benefits of believing the doctor? The patient will have the satisfaction of choosing to believe someone who is an authority and will feel rational and have true, or at least best-informed, beliefs about the world. The downside of believing the doctor is that the belief that one is seriously ill is depressing and frightening. The goal to be well will not necessarily be satisfied, and people don't like to be in a state of goal frustration.

If he instead believes his friend, then he has to suffer the nagging feeling that he is believing someone who doesn't know what she is talking about and that he is being irrational. On the other hand, it allows him to believe that he can be well, which feels very good. The terror will evaporate, a great weight will lift. His goal to be well will, undeservedly, be satisfied. It also allows him to avoid the costly and painful treatment.

If the patient believes his friend, the scientifically minded among us will probably say he is in denial. But sometimes the good feelings associated with believing some fact are greater than the discomfort we get from knowing that, at some level, we have no business believing it. Denial, a defense mechanism that denies the truth of painful thoughts, is a perfect example of this. We get the happiness of believing what we want to believe without the hassle of achieving goals or suffering consequences of painful alternative beliefs. People tend to disbelieve things that threaten their view of the world, even if those things are verifiable facts. So there is a motivation to just believe what you want to be true all of the time—simply believe you are married to your dream partner, and you will feel more satisfied with your spouse!

Unfortunately for all creatures that can suffer, the fuel of evolution is not contentment, but reproduction. Believing things just because you hope they are true is the mind cheating itself, and evolution is probably responsible for making sure there are functions in our mind to keep this from happening—a veridical view of the state of the world, no matter how miserable, has a clear adaptive advantage.

Nonetheless, it's not a perfect system, and people still manage to be in denial about things that others find obviously, painfully wrong. People want to feel good about things, especially themselves and the group they associate themselves with. They want to be successful, healthy, happy, well-liked, attractive, and smart. Thinking highly of ourselves is so pervasive that psychologists have coined at least 15 terms to refer to these "positive illusions." These illusions exist in spite of the mind's valiant attempts to keep denial in check.

One class of these positive illusions makes us believe we are better than other people. When good or bad things happen to a person, there are multiple factors involved. When we are successful, we tend to attribute it to our stable traits (for example, "I'm smart") or to our behaviors (for example, "I got this promotion because I'm a hard worker"). When something bad happens to us, we blame the situation (thinking, for example, "I got fired because my teammates screwed this project up"). This is known as the self-serving bias, which was explored in 1975 by psychologists Dale Miller and Michael Ross. Illusory superiority refers to people's tendency to overestimate their positive qualities and underestimate their negative ones relative to other people. In general, most people think they are above average on most traits and skills (interestingly, for difficult skills, such as riding a unicycle, average people tend to think they are below average). The stronger this bias is, the harder it is for people to find fault with themselves, no matter what happens. It's as though each person has an unfalsifiable theory that no possible observation can refute.

We like information that supports the view of the world we already have. When we encounter such information, we pay more attention to it, remember it better, and use it to build further support for our beliefs. People who read science papers that disagree with their views discount the research (and are less confident in science in general, according to a study by psychologist Geoffrey Munro).[8] The classic example of this is the persistent belief in the full-moon effect, which holds that people act crazily and more violently when there's a full moon. Many nurses and police officers swear that this is true, but every careful study has found that it isn't. But what *does* happen is this: after hearing about the effect, people tend to think of the full-moon effect when there's a full moon *and* people are acting crazy, but not when either one of the two is missing. So when people recall what they think of as relevant information regarding the effect, they only bring to mind the supporting evidence.

This is called "confirmation bias," and once you start looking for it, you see it everywhere. In fact, seeing confirmation bias everywhere is partly due to confirmation bias.

A related bias is "congruence bias," which is like confirmation bias except it affects what information is *sought* rather than how passively received information is treated. People seek information that supports their beliefs rather than evidence that would contradict them.

In one chilling experiment, psychologist Deanna Kuhn looked at how juries determine guilt or innocence. Participants were given some evidence and asked to reach a verdict. You would hope that people would take a look at all the evidence and then decide on a verdict. But instead, participants created a story about what they thought happened, and then sifted through the evidence, looking for information that would support their theory.[9]

For example, if one believed that people who own guns tend to commit murder more than people who don't, the congruence bias would make the person more likely to look at the number of people who own guns and commit murder, rather than the number of people who own guns but don't commit murder, or the people who commit murder but don't own guns.[10]

Is confirmation bias a different effect, at a psychological level, from the congruence bias? Just as the positive illusions might not all be psychologically distinct, congruence and confirmation biases might be different behaviors originating from the same internal psychological mechanism. Scientists want to make names for themselves, so they have an incentive to coin a new term to make it look like they've discovered something new, when in fact they might just be renaming something old. My friend Darren McKee, from the science podcast *The Reality Check,* suggested that this itself is an effect. Like other effects, Darren and I understand that the McKee-Davies bias might be caused by some other underlying positive illusion, leading to an interesting conclusion: The McKee-Davies bias might be the first effect that was created as a result of itself.

I claim that people will be riveted by what they hope or fear is true, but these two emotions are opposites. It would be a weak theory that predicts that we will be compelled by everything. Instead, it predicts that things we feel neutral about will be particularly *un*-compelling. If we look at hope and fear as a continuum,[11] my theory predicts that people care about things on the extremes, but not in the boring middle: an anti–sweet spot.

The two conflicting drives, coming from fear and hope, stem from different psychological needs. Our need to experience fear comes from our need to learn to avoid danger. Hope can help us start difficult or frightening tasks, but the desire to experience hope can also come from bypassing the normal routes to the gratification of being right and feeling good about oneself.

* * *

One of the reasons we like certain works of art is because they remind us, perhaps unconsciously, of real things that we have an interest in. Things that make us hopeful or fearful in real life make for compelling art.

Some of these things involve imagery of elements proven to be beneficial to us in our evolutionary environment. This is most clear in our preference for certain kinds of landscapes in paintings and photographs, and in views from our houses.

Although there is a good deal of abstract and nonrepresentational art, simple landscapes and still lifes are usually preferred by people with no interest in art. In an unscientific but nonetheless fascinating project, artists Vitaly Komar and Alexander Melamid tailor-made paintings for each nation based on the results of survey questions. The surprising outcome was the consistency. People worldwide tended to prefer pretty landscapes (often with people in them) to paintings of other subjects.[12]

It turns out that a lot of what we prefer to see in our landscape paintings can be traced to our survival needs, particularly those of our ancestral evolutionary environment on the African savanna.[13] Psychologist John Balling and educational researcher John Falk ran a study showing that eight-year-old (and younger) children prefer images of the African savanna to any other. After that age, they prefer landscapes similar to where they grew up. Humans spent the majority of their evolutionary history in the African savanna and thus have a genetic predisposition to prefer it, which is overcome by our experiences as we grow up.[14]

The landscape scenes we tend to like, in reality and in depictions, have these elements: a view of water, animal life, and diverse flora, especially flowering and fruiting plants. This has a clear survival advantage, in that it indicates the presence of things to eat and drink.

We also like to be able to see the horizon, presumably because we want to be able to see what's coming. We like low-branching trees, perhaps because they are easiest to climb. We like to look from a place of refuge—where we can see, but are sheltered from the elements and perhaps the envious eyes of others.[15] Balconies and porches are attractive because they provide a view but also offer refuge from the world. Our preferences for how our interiors are designed appear to be influenced by our feelings of safety. A study by psychologists Matthias Spörrle and Jennifer Stich had people arrange bedroom furniture in floor plans. They found that people prefer putting their beds in a place from which they can see the door—but where they are not easily seen *from* the door.[16]

Perhaps most interestingly, we like open spaces with low grasses and with a few trees. Although we might think of grasses cut low in this way as being particularly artificial, research suggests that our concept of nature in its wild state is actually wrong. Most of what we consider nature reserves in the modern world are changed by human beings—we change them to allow certain species (and not others) to survive. There is one reserve in the Netherlands that bucked this trend and operates with as little intervention as possible. On this reserve there are herds of deer and other herbivores, 20 percent of which die every winter. And the grass looks like a putting green from all the eating. Trees are relatively rare because the animals eat them before they can grow tall enough to be recognizable as trees. It looks more like a city park than the idea of nature that we bring to mind— a dark forest.[17] This suggests why we prefer our lawns mowed and like our parks the way we do—they actually represent a natural state that we evolved in. We mow our lawns to mimic what the herds of herbivores we've killed off did for us in the past.

It makes a great deal of sense that we would evolve to prefer landscapes we can prosper in. Even differences between groups of people can predict some of the kinds of landscapes we like. For example, a theory about human hunter-gatherer history suggests that males tended to specialize in hunting and exploration and women in foraging (although both did at least a bit of both). I should note that this idea is hotly debated and by no means universally accepted. Nevertheless, as one would predict from this theory, women indeed prefer pictures with vegetation more than men do, and men prefer pictures featuring a good view more than women do. Preferences can be learned, too. Botanist Elizabeth Lyons used a picture-preference study to find that modern farmers overcome the built-in landscape aesthetics and prefer landscapes that would make for good farmland. Anthropologists Erich Synek and Karl Grammer found that as children grow up they prefer more rugged landscape, presumably because they are better able to traverse it.[18]

We still have deep preferences for nature, even though we are not always aware of it. Walking in a natural environment has even been found to improve cognition, where walking around downtown does not. Research by psychologist John Zelenski has found that walking outside makes you happier than walking inside.[19]

The theory that people like to look at what's good for them predicts that people would like to look at pictures of food. On the face of it, this appears to be false. People do not decorate their walls with paintings of foods they regularly eat. But if we think of trees and animals as food sources, rather than just as things we see when we take walks in the woods, then indeed we do like to look at food, albeit in its natural state. Perhaps cooked food is too evolutionarily recent for us to have a genetic predisposition to like looking at it, which explains why we like pictures of deer but not pictures of roasts. Or, perhaps, prepared food doesn't keep long enough for it to be painted. There is a genre of art called the "game-piece," which shows deceased, hunted animals. Perhaps this is an intermediate form, lying between live animals and hamburgers.

The overall tendency to prefer landscapes in paintings can be interpreted as an unconscious desire to fool ourselves into thinking we're in a beautiful, natural, safe place. This fits into the "hope" theory of this chapter.

How does our desire to attend to and believe things we fear affect our choices of art? Most people don't want scary paintings on their walls. However, people also find themselves compelled to look at scary things when they are encountered, such as fires and car accidents, and they like to watch exciting and scary films.

We can also see the cultural effect of fear in musical preferences. Popular music reflects the fears of the times. A study of popular music from 1955 to 2003 by psychologists Terry Pettijohn II and Donald Sacco Jr. found that during threatening social and economic times, popular music has longer songs, more meaningful content, and is more comforting, more romantic, and slower.[20]

Why do people like sad stories? One reason might be that they make people happy—about their own lives. One study by communications scholar Silvia Knobloch-Westerwick found that people were happier about their own lives after watching a sad movie. It made them count their own blessings.[21]

Why do some people like horror, which, on the face of it, is scary and unpleasant? Our danger bias is a part of it, but we also like horror movies for the same reason we like to play—they are a safe danger, giving us vicarious practice for dangerous situations. Even though the experience of watching a horror movie might be terrifying, we feel, deep down, that we are watching something important, that we are learning something about the dangers of the world we live in; similarly, we can't help but watch a car accident or a fight in the street.[22] The evidence for this is in our dreams.

One current theory of dreams, the *activation-synthesis hypothesis*, holds that they consist of random activation of memories or interpretations thereof. However, dreams occur during rapid eye movement (REM) sleep, which is known to help encode memories of how to do things. More relevant to the present discussion is the *threat simulation theory*.[23] This theory holds that a major function of dreams is to provide virtual practice for threatening events (two-thirds of dreams are threatening). Indeed, people who have experienced trauma have more nightmares. If this theory is correct, and parts of our minds accept what we see in the media as true, then scary movies should provoke nightmares just like real-world life-threatening events do. This in fact happens. After watching five seasons of *The Sopranos* I had anxiety-ridden nightmares about having to deal with violent mafiosi. Nightmares about horror movies are called "nonsense nightmares" because they are training you to deal with problems that don't really exist, such as how to survive a zombie apocalypse.[24]

Interestingly, people get lots of practice surviving zombie apocalypses in computer games. In this case, the game is, as far as many parts of our minds are concerned, real practice. Indeed, a study by psychologists Jayne Gackenbach and Beena Kuruvilla found that high-end gamers experience fewer dreams involving threats.[25] According to this theory of dreams, gaming provides rehearsal for how to appropriately respond to threats.

We know that parts of our minds think scary stories are important because our minds find them important enough to dream about. If scary stories are important, then we are compelled to experience them. "Please watch *The Terminator* again," our lizard brain says, "I need to brush up on how to deal with killer robots from the future."

At the same time, many stories, even harrowing ones, feature happy endings. Danger is more valuable to watch if combined with a way to escape it. There are cultural differences in how stories tend to end. American plays (as of 1927, anyway) feature happy endings more often than those of other cultures, such as Germany's.[26] In spite of some regional differences, however, my theory predicts that, worldwide, stories that can be classified as having happy endings will be more common than stories that can be classified as having unhappy endings.

*   *   *

Some things are compelling in a way that makes us passionate, or give us pleasure. Others are compelling in an addictive, obsessive way: we don't really get a whole lot of pleasure from the activity, but we don't want to stop, either.[27] Experts debate what kinds of things one can become addicted to in a medical sense, but I'm using the term casually, where it can be a positive description of a food or activity that we find so rewarding that we can't help but continue to partake, such as being addicted to potato chips or to a television show, or playing Tetris.

I'm not aware of any scientific research on the subject of what makes a computer game addictive, but I will speculate that we get addicted to games and puzzles because we feel we are getting better at something. Getting better at something makes us feel good about ourselves, fitting into the theme of hope in this chapter.

Grinding is a computer-game term for a situation in which a player must repeat actions over and over again for some kind of (usually in-game) reward. For example, a player might have his character kill monsters, over and over again, to gain powers or status. For some games, grinding is perceived as essential, because meaningful content is expensive to create and consumed far more rapidly by gamers than can be created by game designers. It's hard to make grinding fun, but game designers have found clever ways to make grinding *addictive*. One way to do this is to give a very small reward for each action. In the *Lego Star Wars* games, nearly every action gives some kind of reward. The player feels that he is constantly making some progress, however small. The external reward in the game is matched with an internal reward in the gamer's mind. Often this mechanical reward (e.g., experience points or virtual gold pieces) is accompanied by an intrinsically rewarding stimulus, such as a bright icon and a pleasant sound.

Slot machines are computer games that work by grinding. The player does a very simple action—pulling a lever—over and over again, in hope of financial reward. Unlike *Lego Star Wars*, however, slot machines do not reward every time. Shouldn't they be less addictive?

Actually, no. It turns out that intermittent reward reinforces behavior even more strongly than reliable reward. Some games take advantage of this by rewarding the player at random intervals. In many computer games, for example, *X-Men: Legends,* the player can destroy boxes, but the boxes only occasionally reveal valuables. This encourages the player to addictively destroy every box she sees.

Slot machines have another insidious aspect to them—one can feel like one is getting better at them. To understand this, let's take a familiar example of a child trying to hit a tree with a thrown rock. When the kid throws a rock and it hits the tree, she gets a surge of positive feeling. This is the brain's internal reward system. It rewards all of the ways her muscles moved, so that in the future she will be more likely to hit the tree again. However, if she fails to hit the tree but comes *close* to hitting it, there is still reward, albeit not as much. She perceives that she is getting closer to the target, getting better at the task. *More like that,* the brain says. It feels pretty good to almost hit the tree.

The same thing happens to her when she is an adult using a slot machine, even though in the slot machine case this behavior is irrational. When she gets two "bar" results but not the third, she feels (subconsciously, of course) that she got "close" to the desired result. Her brain assumes she is getting better at the task, so there is a self-generated reward, as Catharine Winstanley found in a brain study.[28] It feels good to get close, even in a slot machine. It is completely irrational because the slots are random, getting two "bar" results is not really "close" to getting three, and of course *how one pulls the bar* has no effect whatsoever on the outcome of the machine. It's completely different from throwing a rock at a tree, but our minds cannot make this distinction.

The vast majority of people at Gamblers Anonymous meetings are not addicted to tabletop blackjack or the roulette wheel—they are addicted to machines. Slot machines and video poker are computer games that take your money even more quickly than the prototypical computer games people think of when they hear the term. What is even worse is that unlike most computer games, casino slot machines collect an incredible amount of data on how the design of their games affects how much money is made. As a result, these gambling computer games are probably the most addictive artifacts ever designed.

*   *   *

The world experts at catering to our hopes and fears are the news media, who created the phrase "if it bleeds, it leads." They know that terrifying stories will get them attention, so they end up reporting as many of them as they can.

Sometimes the news will cleverly play on our hopes and fears one after the other, with a headline or teaser like "IS YOUR CHILD BEING MISTREATED IN PRESCHOOL? FIND OUT WHAT YOU NEED TO KNOW." Like religion, the news scares you and follows up with hope for a solution.

Sometimes the news caters only to hope. No doubt many readers have heard that having a pet increases happiness and health. But few know that the studies reporting no effect are just as numerous. A study showing no such effect is something nobody wants to read. It's not scary nor hopeful enough to grab anyone's attention. It's in the boring dead zone, the anti–sweet spot. As a result, in this case only the positive gets reported.

Like a lot of news, contemporary legends (popularly known as urban legends) tend to be scary. That is because the scary ones are more likely to be retold, as was found in an experiment by psychologists Jean Fox Tree and Mary Susan Weldon.[29] According to my theory, we find cautionary tales compelling because of fear and hope.

*　*　*

While in China, I got acupuncture for some digestive problems. After I was treated my symptoms got worse. I don't feel that I have enough information to know if the acupuncture had any effect at all, good or bad. My acupuncturist, however, had no reservations: she claimed that the Western medicine I was taking (for the same sickness) was interfering with the acupuncture, and, even though she was not a doctor, recommended I stop taking it. I never went back.

After the news, nowhere is the desire to believe what we hope or fear more evident than in the case of bad science, particularly in the realm of medicine. Public superstitions show how easily people get scared. This fear bias results in accepting rumors or hearsay as truth in situations in which people should be skeptical.

For example, people will often believe hearsay regarding disease-causing agents. Polio is a horrendous disease that might have been eliminated once and for all from the face of the earth if not for our fear bias. In areas where the vaccine is distributed, rumors that the polio vaccine *causes* the disease abound, ironically leading to many polio deaths.[30] The folk belief that vaccines are dangerous is not limited to third-world countries struggling with polio.

In the Western world, some people, including some very famous and attractive people (I won't mention names, to protect Jenny McCarthy), believe vaccines cause autism. This belief is not supported by scientific evidence, but people tend to believe it in part because it's frightening. As in the case with polio, it is not the vaccine but the *fear of vaccines* that is the real public health threat. When people are afraid of vaccines, fewer people get them. This means that a higher percentage of the population will be likely to catch an infectious disease, but it's not just those who fail to get the vaccine who are at risk. When a group of people fail to participate, it endangers everybody. But as pediatrician and leading vaccine advocate Paul Offit says, "It's not hard to scare people, but it's extremely difficult to unscare them."[31]

I don't think it's an exaggeration to suggest that the engine of medical quackery is fueled by our desires to believe what we hope or fear. The confirmation bias allows us to notice cures when they work and ignore them when they don't, or indeed, when they make us worse.

Medical quackery is seriously enhanced by the *placebo effect* as well. The placebo effect occurs when the perception that a person is receiving medical treatment positively affects their health and well-being. If a sick person takes a sugar pill (one with no medicinal ingredients), then she is likely to *feel* better. In cases where the problem is psychosomatic, the result of hypochondria, or has a psychological aspect (e.g., problems triggered by stress), the placebo might actually make you better.

Note that there is a placebo effect with real medicine too—because the active ingredient in a drug and the psychological effects of believing you are being treated are independent; the medicine will have its effect and the placebo will have an effect on top of that. The placebo effect is so strong (and, over the years, getting stronger) that placebo-controlled trials are absolutely required in all pharmaceutical research. However, when people evaluate cures in day-to-day life (e.g., my cold goes away whenever I take echinacea), or in unscientific studies (e.g., 70 percent of people who took this snake oil had cold symptoms reduced), the placebo effect, in combination with the confirmation bias, causes people to believe in treatments that don't work, spend money on them, and, worst of all, neglect more effective treatments. Given the strong effects of these biases, it is unwise to put any credibility in anecdotal observations of medicine. If you hear something that you hope or fear is true, treat it with skepticism, because there are forces within you suppressing your doubts.

*   *   *

Like medical quackery, religion too is filled with hope and fear. Christianity, Hinduism, and Islam, in particular, have a great deal to say about what risk one takes by not following their edicts and the benefits to be experienced by following them, both now and in an afterlife.

One thing that can be absolutely terrifying is the thought of dying. Some religions provide some attenuation of this fear by promoting belief in an afterlife. Indeed, even subtly reminding people of death can make them report being more religious and more likely to believe in God, as shown in a study conducted by psychologist Ian Hansen. A study by Chris Jackson and Leslie Francis found that anxious people are more likely to think religion is important than people who have no anxieties.[32]

Dying isn't the only fear that can affect our religious beliefs. Psychological studies have shown that people are more likely to be religious if they feel lonely, have experienced terrorism, or are feeling financial or physical insecurity. Just making an unempowered person more anxious makes him or her a more zealous religious believer.[33]

Religion tends to show up where life is hard. Religion is more common in nations that are dysfunctional, have a lower standard of living (as measured by divorce rates, public health expenditures, doctor-to-population ratios, per capita GDPs, adult literacy rates, and access to safe drinking water), have a higher income inequality, have less trust in general, and are less democratic.[34] Predominately atheist nations tend to be more peaceful than religious ones (as measured by rates of infant mortality, murder, AIDS, abortion, and corruption). People who have not experienced war, have good health, do not feel under threat of terrorism, have stable employment, and are well educated tend to be the least religious. These are the people in least need of reassurance. Finally, religious belief is correlated strongly with hope in individuals.[35] All of this makes it sound like an open-and-shut case. Indeed, many people believe that religion's ability to delude people into a false sense of comfort is the only explanation for religion that is needed.

However, some religions have no completely benevolent gods at all. The religions of the pre-Columbian Mexicans and Mayans were like this.[36] Although the evidence cited above indicates that unstable places are more likely to have religion, this does not *necessarily* mean that the religion they have is of the particularly comforting variety, though it is suggestive. One of the most comforting religions, New Age mysticism, came about in one of the most affluent and secure nations in history. If wish fulfillment were really the only thing going on, you would expect to find particularly comforting religions in the most unpleasant places. Anthropologist Pascal Boyer actually claims that the opposite is true.[37] Perhaps people in dysfunctional environments yearn for some kind of understanding or the feeling that some god is in control. People turn to supernatural agents to understand negative and unpredictable events, rather than blaming these events on chance. Fear makes them turn to religious beliefs.

Psychologist Nicholas Epley ran a study in which people were told that they would end up lonely in life, and they found that this group reported a stronger belief in supernatural agents than a group that was told that they would have a future full of rewarding relationships. People who feel little control of their lives are more likely to believe that the future can be supernaturally predicted. Simply making people feel out of control in the laboratory does things to their minds, making them more likely to perceive patterns in noise, causes between unrelated events, and conspiracies. Psychologist Aaron Kay found that making people anxious makes them more religious.[38]

Cognitive scientist Jesse Bering makes an excellent point that beliefs in hopeful things like the afterlife can't be strictly a function of people being hopeful. Indeed, there are many ideas that would make us happy if we were to believe them, but those ideas are patently absurd.[39] For example, it might make you happy to believe that all the dead batteries in your house will turn into money. Silly, right? Without our natural belief in psychological continuity, the very idea of the mind surviving independently of a living body—the idea of an afterlife—would be considered just as silly.[40]

But there's another fact to consider with regard to religion as a denial of death: not all visions of the afterlife are particularly happy.

Some religions, including some denominations of Christianity, believe in a bad afterlife, often called "hell," for people who do bad
things. The idea of hell is perhaps comforting when thinking about
the bully who reduced your child to tears. This is the common view
of religion as wish fulfillment. However, the idea of hell is not so
comforting when you're scared of being sent there yourself. But
scary things feel important to know about, so the idea of hell is compelling. It might be that hell is particularly compelling for worriers.
Political scientist Daniel Treisman found with surveys that people
who believe in hell tend to worry about a lot of other things too.[41]
Many contemporary Christians tend not to think they are going to
hell. This is not the case with contemporary Muslims, who tend to
think of hell as a real possibility for themselves. As a result, according to anthropologist Lee Ellis, Muslims have even more death anxiety than nonbelievers![42]

Religion can scare people plenty, even without the idea of an
afterlife. Anthropologist Pascal Boyer reports that in places in Mel-
anesia people feel they are under constant threat of witchcraft. Of
course, there are rituals one can do to try to avoid being attacked
by witchcraft, which reduces worry. Sometimes religion brings comfort because it reduces the anxiety the religion creates in the first
place. Examples like the afterlife in Islam cast doubt on wish fulfillment and comfort as a magic bullet, a simple explanation for religion. But that is not to say that it is not often compelling because of
the hope and comfort it brings; it's just that there is more than one
factor at work.

Many readers probably take it for granted that the afterlife is an
important part of religion in general. But this is not true. Many religions have only vaguely defined ideas of what happens to you after
you die, and the nature of the afterlife, if any, has nothing to do with
how you behave in life. However, the most popular religions do have
some kind of afterlife beliefs. Why is this?

Pascal Boyer argues that religion and other beliefs tend to get
interpreted and embellished by different peoples as the religion
spreads.[43] According to Boyer, having scripture, or a doctrine, helps

keep the religion what it is, inhibiting splintering into different sects. For the religion to stay what it is, it needs to be one-size-fits-all, and it needs to be written down to prevent cultural diversity. We call these religions "doctrinal," because they have a specific doctrine. The many religions that do not have doctrine splinter much more often.

However, each geographical area might have its own ideas about gods and spirits. Doctrinal religions demote these to "lesser" gods. The religious leaders of the doctrinal religions are not better at dealing with these local gods than the local shamans, so the religion evolves to pay less attention, relative to primarily non-doctrinal religions, to solving problems of the here and now and more attention to what comes next—the afterlife.

So religions with no doctrine tend to splinter. Vast religions that try to claim dominion over local gods will have fewer believers over time. Without a good claim on the here-and-now gods, nor the afterlife, a religion will appear useless. As a result of all these cultural forces, the most popular religions are doctrinal and focus on some kind of afterlife. So the reason it's easy to think that all religions have an emphasis on an afterlife is because the religions we hear about most would not be the most popular if they didn't! For a religion, belief in an afterlife is *adaptive*.

*   *   *

An interesting thing happens when we start to doubt our views. This doubt causes cognitive dissonance, which is a discomfort as a result of contradictory feelings and thoughts. When people believe in, say, the power of prayer and encounter evidence (be it scientific or from personal experience) that it doesn't work, it causes dissonance. People will act to resolve the dissonance, sometimes irrationally.

A study by psychologists David Gal and Derek Rucker showed that people whose closely held beliefs were undermined engaged in more proselytizing behaviors. Why would this be so? When those around us believe the same things we do, we take this as reason to believe it. So if we can convince the people around us of the belief, then the dissonance is attenuated.[44]

Confirmation bias has a great influence on belief in religion and conspiracy theory. In religion, once people believe in a benevolent God, they start to interpret worldly events differently. People attribute good things that happen to God's will, but not bad things. For example, people might hold God responsible for people being saved from a disaster, but the idea that God might have caused the disaster in the first place might not even cross their minds. Without confirmation bias, peoples' views about God would likely be much more complicated. But because they do not consider bad events (or good things happening to bad people) to be God's responsibility, and perceive good things as coming from God, every good thing that happens supports their theory that God is good.

Likewise, if one's conception of God involves punishment, then bad events get attributed to God as well. For example, when the tsunami struck Thailand in 2004, there were religious people who claimed that it was an act of God to punish Thailand for its wicked ways. Similar attributions to God were made regarding AIDS and gay men.

The fear of divine retribution is tied up with social compellingness as well. Recall Jonathan Haidt and Jesse Bering's idea that our tendency to have religion evolved to help maintain the social order. This requires that the gods have knowledge of us and of what we do. In a major study across many religious groups, historian Raffaele Pettazoni found that the central gods of many religions had a Santa-like intimate knowledge of individuals and what they did.[45] As societies get larger, there is less accountability for your actions—not everybody can know you personally, so reputation means less. As a result, larger societies are more likely to feature religions with gods concerned with human morality, as supported by a study by evolutionary scientists Frans Roes and Michel Raymond.[46] When your fellow citizens can't keep you in line, they have gods step in.

This is how people, in their day-to-day lives, avoid the problem of how an all-good, all-powerful god allows such terrible things to happen. Good things are clearly the result of the goodness of God and bad things are the result of people acting immorally—for instance, Pascal Boyer reports that religions the world over hold that incest can cause natural disasters. As good and bad events are interpreted according to their worldviews, people feel as though they have experienced a great deal of evidence in support of their views. In fact they are counting their own biased perceptions and selective memories as objective observations.

What does it mean to believe that God is "all good?" Naturally, it means that God agrees with whatever *you* believe. And when we change our minds about something, we tend to either think that God changed his mind too or that we have become aware of what his actual opinion was all along. Psychologist Nicholas Epley changed people's minds with persuasive essays and found that people attributed their new beliefs to God.[47]

Boyer relates a description of the Kwaio people and how they interact with invisible spirits, or *adalo,* that reflects the confirmation bias very well as it applies to religious ideas.[48] When something bad happens, such as when a sickness spreads, the Kwaio do divination with a root to get yes or no answers from the *adalo.* They try to find out what the spirit wants as a sacrifice. Often it's pork. If the problem does not go away, they reason that they must have been communicating with the wrong *adalo,* not the one who caused the problem. They repeat this cycle until the problem goes away. Then they reason that the requested sacrifice worked. No matter how many times the sacrifice fails, their belief system provides a way for them to keep on believing in it. Their belief systems are immune to counterevidence. Unless you are Kwaio, this probably sounds pretty silly. But I hope that you can also see how someone might come to find this whole thing rather sensible, given certain presuppositions of how the world works.

Let me bring it closer to home. Similar things can happen in Christianity and other monotheistic religions. If you get a good job, you might thank God for helping you. If the job turns out to feature an abusive boss, then you might reason that in fact God gave you that job to test you and make you stronger. You can fit just about anything that happens into the framework you have. What's alarming about this is that *it will feel like you've found more evidence for the framework itself.*

*   *   *

Confirmation bias is particularly pernicious with conspiracy theorists. Not only do they seek out and give more value to information that supports their theories (the congruence confirmation biases)— as all of us do unless we're being very careful—but they interpret disconfirming evidence as further support of the notion that there's been a cover up. This is one important difference between conspiracy theories and other bad ideas, such as astrology and Reiki: conspiracy theories have embedded within them beliefs about people actively concealing evidence, and thereby are able to incorporate counterevidence into the conspiracy narrative. Any information that appears to disconfirm the conspiracy theorists' explanation of what happened can readily be explained by the part of the theory that describes the cover up.

It is possible that one attraction of conspiracy theories is that they help us make some kind of sense of a chaotic, frightening world. In support of this idea, a study by political psychologist Marina Abalakina-Paap found that people who believe in conspiracy theories tend to feel alienated from society.[49]

*   *   *

We are compelled by things we fear because some parts of our minds treat frightening information as important, be it truth, fiction, or lies. We are compelled by things that give us hope because hope makes us feel happy or better prepared to encounter future difficulties.

3

THE THRILL OF DISCOVERING PATTERNS

We love patterns and repetition. They are pleasurable to experience when they are obvious, and equally enjoyable when they are at first hidden. When we perceive a pattern, the rush of pleasure we feel has been likened to orgasm by psychologist Alison Gopnik.[1] Patterns across space, such as the textures of walls, are important because they often designate contiguous surfaces in the world. Camouflage works because it disrupts the normal visual pattern associated with the surfaces it is on.

Patterns in time, as we experience them in music and in cause-and-effect relationships, are also important. When we notice that one thing follows another with regularity, we learn to make accurate predictions, which is important for staying alive, among other human goals. In the absence of a teacher or some punishment or reward from our environment, we engage in what artificial-intelligence scientists call "unsupervised learning," which is, basically, learning patterns in the environment without any immediate feedback. Much of the time, these predictions are correct. I'm talking about very mundane predictions, such as predicting that your foot will not fall through the floor when you take a step in your bedroom or that the cloud you think you see in the sky really is a cloud. Anytime we recognize anything we are detecting patterns, be it the sound of a dog panting, the way your friend's smile looks, or the taste of milk.

We evolved to be hypersensitive pattern detectors because the cost of missing a pattern (such as a tiger) is greater than the cost of seeing one that isn't there. *Skeptic* magazine editor Michael Shermer calls this "patternicity."[2] Our hypersensitivity manifests itself in a host of psychological biases with names such as the clustering illusion (seeing streaks or clusters in random sequences), pareidolia or apophenia (seeing meaning in random or vague stimuli, such as Jesus's face in toast), and illusory correlation (seeing a relationship that one expects even when no such relationship exists). Fear of autism contributes to people's belief that vaccines cause autism, but the other major factor at work is illusory correlation. Children with autism tend to be diagnosed at about the same age, coincidentally, that they get a lot of their vaccines. First they get vaccines and then they "get" autism. It's a great example of seeing causation where there is only correlation.

I heard a story once of instructor Peter Révész, who teaches the concept of randomness in a brilliant way. He would ask one half of the students to flip a coin 100 times and record the result. He asked the other half to come up with a sequence of heads and tails, 100 long, that *looks* random. He told the groups to write the two strings of heads and tails on the board and that he would come in and guess which one was actually random. With one look he could tell which is the true random sequence—it's the one with long strings of heads or tails.[3]

People underestimate the frequency of long strings of recognizable patterns that inevitably show up in long, random patterns. As a result, when these patterns show up, people think that the underlying causal mechanism that produced the pattern *must not be random at all*. Then they start making guesses about what that non-random process must be.

As we become familiar with a subject, such as a school of painting, or a language, or a musical style, we notice more and more of the patterns that make it up. We can look at this as building a vocabulary of these patterns. When one first hears a new language, one understands nothing and hears only an uninterrupted series of sounds. The sounds might sound different in some ways, but they are in an important sense indistinguishable, because we don't notice what is *importantly* different about them. We can't even tell when one word ends and the next begins. Many times, when it starts raining, there is an arrangement of raindrops visible on the window. Each time it rains the arrangement is different, but they still all look the same (that is, like all the other arrangements of raindrops) because we can't see any patterns in them. There are no patterns there to detect, and our mind becomes blind to them. When observing people, sometimes one will have the impression that members of other races "all look the same," when, of course, all racial groups have a great deal of internal variety. Indeed, people have been shown to be more accurate at distinguishing members of their own ethnic group.[4]

But language *does* have patterns. After many exposures, the sounds of that language, as well as combinations of those sounds, become recognizable. One sentence will sound less and less like all of the other sentences. Before I studied Chinese, I could not tell the difference between Chinese and Japanese. Now I can tell the difference between Mandarin and Cantonese.

Similarly, music might "all sound the same" until we acquire the musical vocabulary to make sense of a new genre. My grandfather once concluded that kids must like contemporary pop music because of its lyrics, on account of the music itself sounds so similar from song to song. Although it seemed incredible to me at the time that he could think that Duran Duran sounded just like Beastie Boys, to him they did. Of course, even fans of popular music often can't tell the differences between different genres of electronic dance music until they get into it a bit. Can you tell the difference between house and trance? How about jungle and dubstep? The differences are obvious to fans.

Patterns also allow for efficient memory storage. For example, memorizing a list of letters, such as "klmrdfkfj," can be challenging. There are nine things you need to remember. However, it gets much easier when those same letters are rearranged as "JFK FDR MLK." If a person knows which famous people those initials refer to, she needs only remember three pieces of information rather than nine. It makes sense that we would have evolved to be happy about discovering things that are easier to remember, because they require less effort. The storage of perceptions of recognizable patterns makes for a kind of data compression in memory. When a pattern is noticed, only a symbol for the pattern need be stored, rather than the many bits that constitute it. In Standard American English, for example, we often pronounce the first syllable of *police* differently from the first of *Pocahontas*. But we hear and notice them as being the same. Similarly, when we look at writing, we learn which marks are extraneous and which are important, allowing us to read the same words in different fonts. People unfamiliar with our alphabet have trouble doing this, because they have not yet learned the essences of the characters.

Perhaps the pleasure we feel when experiencing patterns is a part of what we call beauty. Artificial intelligence researcher Jürgen Schmidhuber's complexity-based theory of beauty holds that things are more beautiful if they require less memory storage because they are simpler.[5] Experts see more patterns than nonexperts. As you might expect, experts are more easily bored with simple stimuli and prefer more originality. Art connoisseurs prefer more abstract and conceptual paintings.[6] Car experts prefer cars with unusual designs. This effect has even been replicated in laboratory experiments by psychologists Claus-Christian Carbon and Helmut Leder—simply exposing people to innovative car designs again and again makes them, in a sense "experts," who then prefer the more innovative designs.[7]

Symmetry, particularly left-right or bilateral symmetry, is compelling because of pattern, and is in fact the strongest predictor of beauty judgments in basic shapes and patterns. Our preference for bilateral symmetry might be explained by the fact that in the natural world, it tends to indicate the presence of a living thing, and gravity dictates that earthbound living things will not have an up-down symmetry, because one side would rest on the ground and another not. Also, when an animal appears as bilaterally symmetric, it is more likely to be facing toward or away from you, because a side view of an animal is not symmetrical.

The detection of symmetry is the detection that something is the same on both sides or all around. It requires ignoring minor differences spread across space. Dyslexia is a disorder characterized by mixing up letters that are the reverse of one another, such as the lowercase *b* and *d*. Dyslexics see these letters as symmetrical and therefore more similar. This doesn't occur to others, and indeed a study by psychologists Thomas Lachmann and Cees van Leeuwen found that dyslexics enjoy the benefit of having better symmetry detection and are also more likely to be able to read upside-down.[8]

Developmentally, it is difficult to learn to make right-left distinctions. We all start out as kind of dyslexic. In the environment of our evolutionary adaptation, this distinction wasn't particularly important. We have to partially unlearn symmetry detection in order to read properly (to distinguish, for example, a lowercase *b* from a lowercase *d*).[9] This suggests that dyslexics would find symmetrical art even more compelling than everyone else.

One incredible finding in psychology is that we are more likely to like and even believe things that we find easy to understand. Before explaining *why* this is true, I'll go over some of the evidence that this *perceptual fluency hypothesis* actually is true.

Things that are easy to understand require less mental processing, which affects the way stimuli are judged. If there is less cognitive processing involved, the idea or experience in question is easier on the mind. Stimuli that are easier to process are perceived as more familiar and visually clearer, more pleasant, louder, longer, more recent, more likely to be chosen, less risky, more attractive, and more truthful.[10]

Statements made in an unfamiliar accent are more difficult to understand and, as we'd predict, people actually believe them less, as shown in a study by psychologists Shiri Lev-Ari and Boaz Keysar.[11] Difficult-to-read text causes a kind of mental speed bump that makes us stop and think more critically, according to an experiment by Jesse Lee Preston. It showed that when reading arguments in favor of capital punishment, people exhibited less of a right- or left-wing bias when the font was difficult to read. People reading about cases in a mock trial were more evenhanded if the text describing the crime had been visually degraded.[12]

People like simple explanations and simple solutions to problems. If ideas are too complicated, people have a tendency to reject them. Psychologist Dean Keith Simonton's study of American presidents and British prime ministers shows that to get elected there is a critical window of intelligence into which the candidate must fall. They can't be too dumb or too smart, and where this window lies is based on the intelligence of the voting population. In America, everyone can vote, so the ideal IQ of presidents tends to be around 119. In the United Kingdom, the House of Commons elects its top leader. Because members of parliament tend to have a higher IQ than the general populace, the window is in a different place. The result is that Britain's prime ministers tend to have higher IQs than American presidents—a 15-point difference.[13]

What makes something easy to process? One predictor is familiarity. The more often we see something, the more familiar it is, and the easier it is to process when we see it in the future. Recall that when we are able to detect patterns, we can efficiently store them into memory. This effect begins even after a single exposure to a particular stimulus—simply having experienced something makes it preferable and more attractive. Repeated exposure to something makes it more prototypical. Prototypical images are the ones we associate most strongly with a given category or word. For example, an equilateral triangle is a more prototypical example of a triangle than one that's two inches wide and a mile high. Predictably, prototypes are considered more attractive. Specifically, people classify prototypical patterns faster and recruit fewer mental resources to perceive them.

Exposure and familiarity even has an effect on what we believe. Experiments by psychologist Ian Maynard Begg show that when people see sentences again and again, they are more likely to rate them as true—even if, when they initially heard the sentence, they were told that it was false.[14]

Of course, there is a limit to how much exposure can increase positive feelings. Eventually we get bored of the same ideas and experiences. We get fond of things, then we get tired of them, and eventually we might want to return to those things.

The proximate reason for the exposure effect seems to be that being exposed to things makes them more fluently processed. But what is the ultimate explanation? It seems that merely experiencing something that we recognize comes with an automatic positive association.

Being able to detect patterns in the world is absolutely crucial for survival. Pattern detection is required for classification, for understanding causal effects, for learning how to act in particular situations. It's no wonder we feel a surge of pleasure when we see a pattern—our minds have evolved to reward us for discovering regularities in the world that might be exploited.

* * *

We evolved to find patterns in the world to help us make predictions, and arts of various kinds all take advantage of this, fooling the mind into thinking it's discovered something about the world, an experience we find pleasing. The easiest example to understand is, perhaps, color. Seeing the same color over and over again in a visual stimulus makes it look better, be it in a website or a match between a woman's eyeliner and blouse color.

Our writing systems are the way they are because the letters are visually distinguishable from each other and because they look natural to us. In fact, the topological properties of writing systems match those of visual scenes from the world. It could be that rather than our brains evolving for writing systems, our writing systems evolved to fit the brain, as a study by neuroscientist Mark Changizi suggests.[15] As the shapes we see in letters are common in the natural world, it's easier for us to process them. As a result, they are pleasing to look at. As evidenced in the popularity of calligraphy, in many cultures, across many eras, we find letter forms beautiful.

* * *

A possible reason narratives are compelling is that they make sense according to the structures we use to understand the world. Take, for example, the sentence "The king died and the queen died." Compare that to "The king died and then the queen died of grief."[16] Although the second one is more complicated, in that it has more words and information, in another way it is less complicated. The events in the latter sentence are not disconnected but fit together in a way we can understand. Because it fits into our notions of how people behave, we can almost predict the second event from the first, making the whole thing easier to remember. Of course, another reason the second sentence is more memorable is because it contains social causes.

Writers recognize that repeating symbols draw the reader in. In the world of theatrical improvisation, satisfying endings often "reincorporate" elements from the beginning of a scene.[17]

Foreshadowing is a way of using repetition to interest the audience. Sometimes foreshadowing is obvious. In film this is called a "leave-behind." Literature sometimes refers to it as "Chekhov's gun," because the playwright Anton Chekhov wrote that an artist "must not put a loaded rifle on the stage if no one is thinking of firing it." Symbolism also can take advantage of repetition. When we first encounter the paperweight in George Orwell's *1984,* we might not pay attention to that. But when the Thought Police arrest Winston Smith, it shatters on the floor. The power of the symbol is evoked by the memory of the role of the paperweight earlier in the novel.[18]

We can also find some things people say very compelling. People talk all day, but only some utterances are repeated and turned into quotations and idioms, while others fade from our cultural memory. Why?

Certain quotations and idioms are sticky; they have a ring to them that catches in our memories or appear to say things that strike us as true. Aesthetic principles that govern the appreciation of other arts also apply to quotations. We love symmetry (e.g., "love the way you want to be loved"); incongruity ("keep your friends close and your enemies closer" or "the bigger they are, the harder they fall"); and repetition, be it in rhyme ("see you later, alligator"), alliteration ("a dime a dozen"), or word structure ("a penny saved is a penny earned").

A shocking number of famous quotations were never even said. What happens is that somebody, usually somebody already famous, says something and the quotation gets altered to be more memorable, more succinct, more compelling. For example, Albert Einstein is famous for having said of quantum mechanics "God does not play dice." Actually, what he said was the much more cumbersome "It seems hard to sneak a look at God's cards. But that he plays dice and uses 'telepathic' methods . . . is something that I cannot believe for a single moment." Doesn't have the same ring, does it?

Titles of books are much shorter than they used to be. Short book titles correlate with the growth of the publishing industry, perhaps because short titles are easier to market in a crowded marketplace. This was found based on a study of 7,000 books in eighteenth- and nineteenth-century England by literary scholar Franco Moretti.[19] Short titles are easier to digest, and long titles are so unusual now that they sound humorous (e.g., *A Voyage to the South Sea, Undertaken by Command of His Majesty, For the Purpose of Conveying the Bread-Fruit Tree to the West Indies, In His Majesty's Ship the Bounty, Commanded by Lieutenant William Bligh* was the title of a book describing the mutiny on the *Bounty*). The actual title of what most people consider *The Origin of Species* is *On the Origin of Species by Means of Natural Selection, or the Preservation of Favoured Races in the Struggle for Life.* Just as we make quotations shorter, we seem to do it with titles too. Why? Because ease of processing often increases preference and shorter titles are easier to process.

The value of repeated elements is strikingly obvious in music. Music scholar David Huron conducted a study finding that 94 percent of musical passages are repeated elsewhere in the work.[20] Musical repetition works on at least six different levels of abstraction. Perhaps the simplest pattern is the use of the same instruments throughout a piece.

A rhythm can't even exist without repetition. Beats have time between them, and a constant beat is a repetition of the amount of time between beats. Even a gradually accelerating rhythm has its own kind of pattern, as the times grow linearly or otherwise.

At another level are musical motifs, which are repeated elements, usually melodic. In his operas, Richard Wagner used a motif for a particular character again and again. This technique is used to great effect today in film scoring.

Music with lyrics employs the same kinds of elements as described above with poetry. In addition, lyrics match the rhythm in music. In fact people subconsciously choose words that rhythmically fit with what they are saying even in normal speech.[21]

Verses and choruses also repeat, allowing us to become familiar with them over the course of a piece of music. And of course we all have had the experience of appreciating a piece of music more and more as we hear it.

*   *   *

Music creates patterns over time to capture our attention. But even little bits of text use repetition to stick in our memory. As in poetry and song lyrics, compelling idioms and quotations often make use of rhyme, such as "birds of a feather flock together."

They don't just sound better; people actually find idioms that rhyme more believable. For example, a study by psychologists Matthew McGlone and Jessica Tofighbakhsh found that people will judge "woes unite foes" to be a more accurate description than "woes unite enemies" because the rhyme increases processing fluency.[22] Political slogans, too, sometimes use the same phonetic tricks that poets use, as in the wildly successful "I like Ike." This has cleverly been called the "rhyme is reason effect." Why is it clever? Because we're probably more likely to believe in the effect when the effect has such a great name. It has a pattern in that it reminds us of the saying "no rhyme or reason" and the two content words start with an *r* (that is, *are*) sound.[23] But it's a little bit incongruous in that it differs slightly from "rhyme or reason," piquing our interest.

*   *   *

Repetition also helps explain our love of sports. If we look at the rules, we can see that they define a structure that includes a good deal of repetition. In hockey, a puck is dropped between two play-ers. In soccer, when a ball goes out of bounds players throw it back in with their hands. In basketball, there are foul shots. These events occur over and over in games. These repeated behaviors are familiar to spectators, who simultaneously enjoy the ritual and rejoice in the nuances of how they play out each time.

* * *

Because finding patterns in the world is so important, we evolved to experience finding a pattern pleasurable. This is the ultimate ex-planation. But what is the proximate explanation? The brain uses a chemical called dopamine to recognize patterns. Dopamine is a neurotransmitter, one of the chemicals that the neurons in your head use to communicate with each other. Not only does it matter which neurons are passing information, but it also matters *how* they trans-mit it—that is, which neurotransmitters actually carry the message. Pattern recognition is one of the many functions of this fascinating chemical.

When the dopamine pattern-recognition system is working, you get high-quality beliefs out of it. However, we've all had the experi-ence of seeing a pattern that is not there—a figure in the shadows or a face in the clouds. Dopamine is the neurotransmitter that tags perceptions as meaningful, prompting us to make sense of them.

It is likely that the sensitivity of our pattern detector is in part de-termined by our dopamine levels. High dopamine makes everything look significant. People suffering from schizophrenia overproduce dopamine, and see patterns even in random noise and notice even the loosest of connections between ideas.[24] Research psychiatrist Shitij Kapur theorizes that when we think something is important, our brain struggles to rationalize an explanation for it. He calls this "biased inductive logic." We try to figure out why the stimuli feel so important. As a result, the explanations are culture specific (it's the doing of the FBI, a curse from a witch doctor, etc.).[25] This is profound. It's not that a schizophrenic patient reasons that the FBI is controlling his or her mind and then the feeling follows. The feeling comes first, unconsciously, feeding up from the dopaminergic system. Only then does the rational, deliberate part of the mind concoct some story to explain the feeling.

But you don't have to have schizophrenia to see patterns where none exist. People who naturally have more dopamine are better at seeing faces in jumbled images and are also more prone to see faces that are not there.

People who believe in the paranormal tend to be bad judges of randomness. Suppose you roll a die three times, and you get 5, 1, and then 3. Suppose someone else rolls a die three times and gets 2, 2, and 2. Both results (in those specific orders) are equally probable, because the dice rolls are independent. However, a study by psychologist Peter Brugger found that believers in the paranormal are more likely than skeptics to think that patterns like 2–2–2 are less likely to occur.[26]

Does dopamine *cause* the pattern finding? In another study Brugger found that skeptical people tend to have lower dopamine levels and those who believe in the paranormal have higher dopamine levels. In fact, artificially increasing dopamine in skeptics (in pill form) starts making them see more patterns, including nonexistent ones. The dopamine level thus causes the effect, and not vice versa.[27]

Superstitious learning is an effect discovered by behaviorists in psychology. You can train rats and pigeons to press a bar to get food. This is called "reinforcement." However, if you give them food at random intervals, a very interesting thing happens: the animal will start displaying odd, idiosyncratic behaviors to get the food. Here's how it works: suppose a hungry rat happens to be scratching its right ear just before the food arrives. The rat's mind is desperately trying to figure out some pattern to the food arrival. It assumes (unconsciously, probably) that the scratching led to the arrival. So it keeps scratching. Such animals are acting as though they believe that some combination of what they did resulted in food delivery. They are inaccurately inferring cause-and-effect relationships. If you test ten rats, you'll get them all doing different superstitious behaviors, based on whatever behavior they happened to have been doing before the food came.

Normally, the pattern-detection mechanism in our minds is anchored in reality—that is, we tend to perceive patterns that are real, in whatever sense patterns can be considered real. But especially under conditions where the link with reality is compromised—such as when the subject is high on drugs, dreaming, under high stress, or under sensory deprivation—the system starts making wacky conclusions about the world and seeing random events as meaningful and related to each other. Deny the pattern detector its normal input and it starts trying to make sense of essentially random or missing inputs: a person might stare at a television on channel fuzz and try to make out a picture, or they might make mountains out of molehills in the dark, misinterpreting the random firing of the sensors in their eyes.

We all differ in the number of hippocampal neurons we have, and that number might be related to our ability to see patterns. Perhaps both our dopamine levels and our hippocampus sizes predict pattern detection and religiosity.

Based on this, I conjecture that religious and paranormal believers are more likely to get schizophrenia, which is characterized by an excess of dopamine, and skeptics are more likely to get Parkinson's disease, which is characterized by a *shortage* of dopamine.

Schizophrenia has a genetic component. A milder expression of these genes can result in a condition called "schizotypy," which is characterized by social withdrawal, odd behavior and thinking, magical thinking, and occasional quasi-psychotic episodes with hallucinations and delusionlike ideas. The hallucinations are milder—more like strange perceptual experiences, such as seeing spirits in a meadow.

Why hasn't evolution removed these genes from the population? The replication of the gene for schizophrenia-spectrum disorders might be due to the fact that in many traditional societies, schizotypals or epileptics are often the witch doctors and shamans, or they are perceived as blessed and are more likely to set the religious tone of the society.[28] Schizotypy is more common in religious people and cult members than in the general population, although it bears noting that some research finds psychotics in general tend to be less religious (schizophrenia is a kind of psychosis). Research by psychologists Rachel Pechey and Peter Halligan shows that people who experience delusions tend to have more paranormal beliefs.[29] As anthropologist Alfred Kroeber says: "The shaman displays his possession by a spirit by publicly re-enacting his specific personal experience, that of a man suffering from a particular mental affliction. His projections, his hallucinations, his journey through space and time, thus became a dramatic ritual and served as the prototype for all future concepts for the religious road to perfection."[30] Schizotypals also tend to have very concrete religious beliefs. Most shamans are not celibate, which allows their genes to spread. This suggests not only that schizophrenia-spectrum effects might account for some of people's continuing beliefs in religion, but that schizotypals might even have originated many of the beliefs religious people hold in culture today.[31] The relatives of schizophrenics tend to be more creative and more religious, replicating the genes in the population without displaying full-blown schizophrenia.[32]

Schizotypalism affects about 3 percent of the population,[33] but the problem of perceiving nonexistent patterns is so widespread that it goes by several names in different disciplines. In statistics, claiming that you've found a pattern when there isn't one is called a type 1 error. The literature on psychological biases describes the finding of clusters or streaks in random data, such as basketball shots and coin tosses, which is called the "clustering illusion" or "illusory correlation." A related effect is subjective validation, a bias that makes someone more likely to see a relationship between two events because such a relationship would have personal meaning to the perceiver. Just because different scholars have come up with different terms does not mean that they are all separate effects. It could be that all these observed effects are caused by the same underlying psychological and brain mechanism. If this is true, then dopamine levels and hippocampus size (as caused by many factors) are good candidates for what this underlying mechanism might be.

Could beliefs in the paranormal and many religious beliefs be linked to an overly amped-up sensitivity for pattern detection? It is likely. In addition to the evidence above about schizotypals, psychologist Paul Rogers found that people who believe in the supernatural are also more likely to be susceptible to the conjunction fallacy (believing, mistakenly, that conjunctions of facts are more probable than the individual facts themselves, such as believing that someone is more likely to be a Republican banker than a banker), to have poor reasoning with probabilities, and to have more misperceptions of randomness.[34] That is, people who believe in such things also have reasoning problems in their perceptions and interpretations of other, more mundane aspects of the world.

Skeptics are more likely to miss patterns that really are there, where believers might see too many patterns.

Problems with overactive pattern detection can affect religion, but pattern helps religion maintain itself in a more obvious way—by repetition of rituals and communications of the core belief systems. Many religions require weekly or even daily readings from a sacred text or attendance at clergy sermons. The themes and beliefs of these religions are heard over and over by the congregations. This repetition has obvious implications for ease of processing and memory, ultimately facilitating belief.

People tend to look for explanations for the unusual or unexpected, particularly misfortune. As noted earlier, religions the world over explain misfortune as being the result of punishment or retribution from gods or witches.[35] This is particularly likely if several misfortunes occur at once. Because we are bad at detecting randomness, and think we see patterns that are not there, our minds crave an explanation.

* * *

There is good reason to think that psychological factors influence the belief (and disbelief) in conspiracy theories. A study by Viren Swami found that people who believe in one conspiracy theory tend to believe in others.[36] This suggests that there can be something about you, some trait, that makes you prone to conspiracy theorizing. Or, for nonbelievers, prone to skepticism.

People of lower social status should be more prone to conspiracy theories for two reasons: first, they are probably less privy in their day-to-day lives to knowledge of how power actually works in the world. This not only leaves a gap of knowledge that the mind wants to fill but also makes the workings of society appear chaotic, which encourages superstitious reasoning. And recall that people primed to feel out of control are more likely to see patterns in random stimuli.

Secret knowledge of how the world works would also be more valuable, as such people have further to go up the social ladder than a higher-status person. Indeed, people who endorse conspiracy theories are especially likely to feel angry, mistrustful, alienated from society, and as if larger forces are controlling their lives. They feel out of control.

Conspiracy theorists are also often immune to any kind of counterevidence. As we've seen, when they encounter contradictory evidence, they don't see it as evidence that disproves their theory, but as further evidence of the cover-up. People are creative and can figure out how to fit new information into their conspiracy theory narratives. Ironically, the smarter a person is, the better they are at it! For example, a conspiracy theorist might believe that aliens have visited earth and that the United States government is covering it up. If the government releases information explaining that a particular UFO sighting was a secret aircraft owned by the military, conspiracy theorists will classify that press release as a cover-up. As people classify things according to their theories, it feels to them that the theories are gaining evidence (religious people also feel they have experienced a great deal of evidence for God's existence as a result of classifying so many day-to-day things as being divine interventions). When evidence and counterevidence both count as evidence for the theorist, it's easy to see how everything they experience reinforces the explanations they already have. Talking to a conspiracy theorist can be like playing table tennis in a room with an oscillating fan. A bit of advice to the reader: when debating with a conspiracy theorist, don't debate evidence; they will have an answer for everything. Debate their epistemology—try to show them that they have set up a way of thinking that is immune to any *conceivable* counterevidence. They are trapped in a vicious circle that's very hard to escape.

As evidence against the conspiracy theory grows, the theory itself becomes more complex to account for the growing reasons to disbelieve it. Conspiracy theories are complex and constantly changing. When scientific theories develop ad hoc additions to account for any state of the world, it is seen as a problem. The theory loses parsimony and it loses ground to other, simpler theories that explain the same observations. Unfortunately, this does not seem to happen with conspiracy theories.

As the complexity of the conspiracy grows, theorists are ever *more* fascinated. They can feel this fascination even if they don't believe in the theory. For many, the wild and unlikely theory that the government constructed HIV in order to kill gays and blacks is far more fascinating than how HIV actually became a problem. The bigger and more convoluted the conspiracy theory is, the more important it feels.

A mistrust of authority, what Michael Shermer calls the "heretic personality," is a hallmark of the conspiracy theorist.[37] So deep is this mistrust that many are willing to believe contradictory views, as long as both of those views are different from what the authorities are telling the people. An interesting study by psychologist Michael Wood showed that if someone believes that Princess Diana faked her death, they are more likely to also believe that she was murdered.[38] If people are willing to entertain contradictions in their own theories, how can counterevidence have any chance at all of changing their minds when it's presented?

This is one area in which religious people and conspiracy theorists differ—the religious tend to trust authority more than the average person.[39] In one experiment by psychologist Daniel Wisneski the more religious participants tended to trust the ability of the Supreme Court to make good decisions whereas participants with strong moral, but not religious, convictions felt distrust.

* * *

Ideas that seem absurd at first become more plausible after repeated exposure. When an idea gets reported repeatedly in the media, or at church, or by the people you socialize with, it has an effect of making the idea seem more believable. Why does the "availability cascade" happen? I suspect there are two processes at work here. The first is the "bandwagon effect," which makes people do and believe things simply because other people are believing and doing them. The reason most relevant to this chapter, of course, is that repeated exposure creates a pattern, increasing cognitive fluency, resulting in a favorable evaluation. We simply like the idea more. When a person repeatedly hears about an idea, there is social pressure to accept the idea (we don't want to challenge our peers) and there is a cognitive pressure to accept the idea (common ideas are just easier on the mind). What might appear unusual and weird at first becomes familiar, softening our minds to the idea.

The availability cascade can lead to good or bad outcomes, depending on the quality of the idea. As an example of a bad outcome, let's look at the frequency of child molestation in day care centers and compare it to the terror felt by parents. During the 1980s and early 1990s, people started to be very frightened that their children would be sexually assaulted in day care centers. Some well-publicized examples made the problem seem more common than it really was—a study at the time found that kids were actually in more danger in their own homes than at day cares.[40] This fear still lingers today.

One study in the United Kingdom found that just 2 percent of caregivers for children under five are men.[41] An anticipated fear that male childcare workers are likely to be child molesters has likely had a crippling effect on men's abilities to get jobs as childcare workers. Is there anything wrong with this? There might be. Oxytocin is a hormone that causes bonding between people. When mothers cuddle their children, they get a boost of oxytocin. Men, relative to women, get the boost when engaging in more rough-and-tumble play. The theory is that men and women evolved to provide the offspring with different experiences. With only women in day care centers, society seems to have inadvertently made the decision that we don't need men to help raise children outside of the home. It is time we asked ourselves if this is really what we want.

As conspiracy theories and ideas of alien abduction get more popular, they become accessible narratives in people's minds. Then, when people have new experiences or hear new stories, those narratives are there for people to potentially fit the new information into.

Let's look at the common view of alien abduction. The idea that aliens are visiting our world and abducting people has been raised over and over in Western culture. When you hear about alien abduction there's a part of your mind that starts to think there's something to all the hype. In the media, as in rumor, this can be a self-strengthening cycle. When one seeks out a group of other believers, for example, proclaimed abductees, one becomes part of a subculture where the availability cascade becomes even more intense, as the ideas get recycled with more regularity and less counterpoint.

* * *

It's obvious how the availability cascade works with religion. People who grow up in a culture where nobody questions the existence of gods and spirits rarely question their existence. This is true for religious beliefs, as well as for interpreting events (e.g., pregnancies, deaths, temporal lobe seizures) as having religious significance.

One morning I was lying in bed and out of the corner of my eye I perceived a shadowy thing moving by my window. I looked over and saw nothing. I don't believe that there are spirits that can manifest themselves in this world. I do believe that the peripheral areas of the eye are sensitive to movement and that perception is sometimes subject to random fluctuations that can result in misperceptions—which is exactly how I interpreted the situation. My "perception" of something moving was, in my interpretation, a glitch in my visual system, not a spirit.

However, people who believe that we are surrounded by spirits we can sometimes see might have interpreted this experience very differently. Given their beliefs, it would be rational to assume that this could very well have been the experience of a spirit.

I met a scientist once who believed in magic. When I asked him why, he shook his head at me, incredulous, and said, "Just open your eyes." He saw magic happening every day. Believers and nonbelievers alike have open eyes. What's different is how they understand what they see. We *interpret* our experiences in a way that makes sense given our beliefs about the world. He and I both remember our experiences as being further evidence of the very belief systems that generated them. The problem, in both cases, is that we never got any outside confirmation that our interpretation was correct. The result is that two people with the same perceptual input end up remembering them as confirmations of completely different world views. It's no wonder it's so hard to change people's minds, skeptics and believers alike.

Over the years, the "understanding" becomes unconscious. It *feels* like we are perceiving something directly. For example, in a particular culture someone might fall into a trance. The people who witness this event feel that they are witnessing a possession by a spirit. To them, it's as obvious as when we see someone eating enthusiastically and "see" that someone is hungry, or "see" that a child really likes his ice cream. These are assumptions at an unconscious level that feel just like regular perceptions.

Often patterns are hard to see, and we don't perceive them at all unless we're looking for them. We can deliberately change how sensitive we are to patterns. Cognitive scientist Jesse Bering puts it well:

> When people ask God to give them a sign, they're often at a standstill, a fork in the road, paralyzed in a critical moment of existential ambivalence. In such cases, our ears are pricked, our eyes widened, our thoughts ruminating on a particular problem. God doesn't tell us the answers directly, of course. There's no nod to the left, no telling elbow poke in our side or "psst" in our ear. Rather, we envision God, and other entities like him, as encrypting strategic information in an almost infinite array of natural events: the prognostic stopping of a clock at a certain hour and time; the sudden shrieking of a hawk; an embarrassing blemish on our nose appearing on the eve of an important interview; a choice parking spot opening up at a crowded mall just as we pull around; an interesting stranger sitting next to us on a plane. The possibilities are endless. When the emotional climate is just

right, there's hardly a shape or form that "evidence" cannot assume. Our minds make meaning by disambiguating the meaningless.[42]

In cultures where there are differences of opinion regarding religious matters, there are competing cascades. Public opinion can rapidly change if at least 10 percent of the population actively proselytizes, as a simulation conducted by Jerry Xie showed.[43] It is a common cult strategy to forbid members from communicating with anyone outside the cult. This turns the normal availability cascade into a prison for the mind.[44] The religion's leaders might know explicitly that the taboo of talking to those outside the cult will help the cult survive, in that talking to nonbelievers might undermine faith. The members of the religion might also know it implicitly, in that they might not be aware of the function of such rules, but the religion holds them nonetheless because it was an adaptive trait that helped it compete with other religions as it evolved. Many religions can be successful without this strategy, which raises the question of why the taboo works well for some religions but not others.

With the Internet, it's easy to find people who agree with just about *any* belief. People who disagree with us can be annoying and there's a temptation to not talk to them—it doesn't matter if the subject is religion, politics, child-rearing practices, or whether the *Star Wars* prequels were any good. But if you decide to stop talking to people who disagree with you, know that it's hard to ever change your views, even if you should.

Patterns also make religion attractive because of the frequency of patterns of behavior: rituals. The repetition of actions, including recitation of meaningful text, reinforces the familiarity (and thus, the believability) of the religion's doctrines. In fact, repetition in a ritual increases its perceived effectiveness. Even nonbelievers derive comfort from ritual, perhaps because it gives them a feeling of control, as psychologists Michael Norton and Francesca Gino found.[45] Of course, too much repetition leads to boredom. This "tedium effect" has been reported by anthropologist Harvey Whitehouse as a problem for missionaries trying to evangelize.[46] As we will see in the next chapter, as alluring as patterns are, some incongruities are necessary to keep us riveted.

One common explanation for why religion exists at all is that it provides explanations for phenomena we don't understand. Anthropologist Pascal Boyer believes that many religious beliefs are attempts at explaining our own intuitions. Because much of our mind's processing is hidden from our consciousness, we feel that ideas and feelings just pop into our heads from nowhere.[47] This is true for moral intuitions as well as artistic ideas. Such things can be more easily understood if we believe they are coming from some outside source, such as a god or a muse.

When we hear stories, be they religious or otherwise, they can be hard to remember if they contain too many unfamiliar elements. Our minds tend to replace some of them (for example, a canoe) with a more familiar one (a rowboat) as found in a classic study by psychologist Frederic Bartlett.[48] Jokes about old presidents get changed to jokes about new presidents. As people revise ideas to better suit their cultures, and what they are familiar with, the ideas are more likely to be remembered and passed on to other people.

*   *   *

Supernatural agents tend to fall into certain types because they fit patterns we understand well. Compelling supernatural agents align with the institutions we already have of objects, animals, and people, usually with only one or two characteristics from another category. We have core theories of knowledge about the world—physical, social, and perhaps biological, and many compelling beliefs are the result of mixing the intuitions from them, such as applying beliefs from social knowledge to the physical world. Examples of this mixing include ghosts, a tree that can hear you, or a statue that bleeds. Such ideas are more common in religion than, for example, a god who only exists three days out of the week, because that is something that neither tools, animals, people, nor objects tend to do. Studies have shown that the best candidates for religious supernatural agents are those that are clearly in one category, but with only one or maybe two characteristics from a different category. More just muddies the waters. These strange but yet fairly uncomplicated ideas, called "minimally counterintuitive," are more interesting, memorable, and easy to communicate.[49] They're riveting.

Some ideas are "mundane concepts." They are so normal sounding as to just be plain believable—and forgettable. If someone were to say "I know a woman named Vanessa with brown hair," nobody would ever have reason to doubt this. On the other hand, they also would not find this fact particularly memorable or interesting.

Somewhat more compelling are what can be called "*merely counterfactual concepts*," which are probably not real, such as "a newborn baby who can cook beans." This is somewhat arresting, and a better candidate for a religious belief than a mundane concept. But it is not quite as good as a minimally counterintuitive idea because none of the characteristics cross category boundaries. In other words it *just happens to be true* that babies can't cook beans, but it's not that babies are in a general category of things that can't. Babies are people, and people can cook beans, so it's not as severe a violation, as, say, a crying statue would be.

The minimally counterintuitive concept of a man who can walk on water is in the sweet spot—it is exactly the kind of idea that makes a religion thrive. It involves enough of what we understand to make it comprehensible, in its own way, but violates what we know to be true in the real world enough to make our minds continue to work on it, to try to figure it out. Like a paradox, it nags at our attention for a long time. We can't find the closure that would allow us to neatly categorize it, and then forget about it, as we do with many mundane concepts.[50]

Whereas mundane and counterfactual ideas do not go far enough, "bizarre concepts" go too far. For example, a man who can walk on water, give birth, read peoples' minds, and create glasses of milk through force of will alone sounds outlandish, and people are less likely to accept that it could be real.[51]

Likewise, some human beings are thought to be special. Perhaps they can predict the future, or have lived for 200 years. To understand these cases (and I'm talking now about understanding, not particularly believing) someone uses the normal mental template of "person" and changes it. Usually there are only one or two changes. Our "ontology" is our understanding of the basic kinds of things in the world (for example, plants, animals, rocks). A counterontological idea violates what we know and this incongruity draws our attention. Indeed, cross-cultural psychological experiments with novel supernatural concepts support this. Ontological violations were remembered better than mere oddities. A man who can walk through walls is better than a man with six fingers, because the latter does not violate a man's perceived essence. So although surface appearances can change greatly from culture to culture (this god has horns, that one has three legs) the ontological violations tend to be similar, falling into a few kinds: people or animals that violate physical properties (e.g., incorporeal or invisible), biological properties (e.g., don't die, result from a virgin birth), or psychological properties (e.g., have supernaturally extended perception). Similarly, objects can have biological properties (such as bleeding) or psychological ones (such as being able to hear). If there is more than one violation, it is more easily forgotten. Multiple violations are rare and tend to be limited to literate theology and are not popular with the public. Psychologists Michael Kelly and Frank Keil studied tales from Ovid and the ancient Greeks and found that mythological and fairy tale metamorphoses tend to happen between kinds of things that are similar.[52] For example, in mythology it is more common for people to turn into animals than into plants. And, consistent with the social compellingness hypothesis, the nature of the changed person's mind tends to remain unaltered. Counterontological beliefs, in small quantities, are more easily remembered, giving them a distinct advantage for cultural survival.

The vast majority of religions have something to say about death. Pascal Boyer believes this comes from counterontological thoughts about dead bodies. When a person dies, especially someone close to us, we have conflicting thoughts. The sight of their body still evokes some of those same ideas of affection, including the person having a mind and our feelings of love for them, yet now our biological intuitions are telling us that the body is an *object*, and our contagion systems are telling us that the body is dangerous! It is common for religions to hold that a corpse is polluted, or impure. Our minds are torn—it's a dead animal and at the same time it feels like it's still a person with a mind. These conflicting thoughts and emotions require some kind of resolution, and religion is up to the task.

Just as we have "object permanence" processes that help us keep track of objects when they leave our field of view, we have "person permanence" processes that allow us to understand that people do not vanish from our world when they walk out of the room. When a loved one dies, our theory of mind and person permanence processes "believe" that the person still has a mind. However, it feels obvious from looking at a dead body that the mind is not there. Religious thinking steps in and generates a belief that the mind has escaped the body, because it *feels like the mind is still there*. Indeed, we can still feel anger or love directed at dead people. Because the dead body brings up these strange feelings and thoughts, all religions hold that something must be done with the body.

Boyer's counterontology theory predicts why *all* religions have things to say about the dead and why spirits of the dead are the most widespread kind of supernatural agents the world over. Cultural inventions such as a statue that cries or a person who is descended from a god are compelling because they cause dissociations with our intuitive systems. Dead bodies are naturally occurring objects that automatically cause counterontology.

After a while, our intuitive minds get used to the idea that the person is actually gone, leaving us only our memories. This is the explanation Boyer provides for "double funerals," which are very common the world over. In these cases, a body is treated (buried, burned) immediately (for safety reasons, ultimately). Months later the body is unearthed, and the bones are put in their final resting place. For Boyer, these two rituals mark psychological changes in our feelings about the deceased persons' bodies. The first ritual is for our feeling that the person is not really gone and the second marks the fading of that feeling.

*   *   *

These counterontological thoughts and feelings about bodies explain our fascination and fear with varieties of living and dead people. Boyer says that peoples' basic concepts relevant to religious agents are life, sentience, and personal identity; in Justin Barrett's version of the theory people think in terms of psychology, biology, and physics. I suggest that our intuitive notions of people involve four basic components: body, life, animacy, and mind.

My "theory of the living dead" is a speculative description of our fascination with different combinations of these fundamental person concepts. Following Boyer's counterontology theory, conflicts in intuition should create compelling concepts. You can chart out all the combinations, making 16 rows, of these ontologies. Having a "body" is just what it sounds like—knowledge or perception of a physical body. "Life" is a concept we get from intuitive biology. We experience feelings, at a visceral level, regarding whether a biological object is alive or dead. "Animacy" just means that the body moves on its own volition (as opposed to a rock rolling down a hill). "Mind" means that our theory of mind understands the being to have things like beliefs, goals, perceptions, and feelings.

By looking at every combination of these concepts, we can find compelling supernatural agents from fiction and religion and also combinations that are curiously empty. I will describe a few entries in this grid.

What has life but no body, animacy, or mind? Perhaps some kind of life force. The doctrine of vitalism, now discredited by science, holds that things are alive because they have some kind of life energy. East Asian philosophy posits the existence of *qi* (Chinese; *ki* in Japanese). The idea of a life force is a part of natural psychology and many religions.

When a person is sleeping or in a persistent vegetative state, they have a body and life, but not animacy or a mind. Because we are familiar with sleeping, I believe that this particular arrangement is not as compelling or frightening as the others. This observation would predict that people would find a persistent vegetative state less interesting and creepy than either locked-in syndrome or severe mental illness. (Locked-in syndrome is when a person is fully conscious but unable to control their body. These people have a body, life, and a mind, but are missing animacy. A tree that can hear conversations also fits this description.) I cannot think of something that has animacy without any of the other concepts.

The idea of a zombie includes the concepts of a body and animacy, but not life or a mind. This makes zombies appear fundamentally weird, and fascinating. Zombies are popular in speculative fiction, but the basic idea originated in Haitian religion. Sleepwalkers are also interesting, in that they have a body and animacy, but are apparently mindless.

Many other combinations of these characteristics have parallels in folklore, myth, and fiction.

However, just as single ideas cannot be too simple or too complex in order to be acceptable, larger narratives also must fit into a sweet spot. Religious stories as well as fantasy fiction tend to have mostly mundane concepts, with occasional minimally counterintuitive concepts thrown in to keep them interesting. Studies have shown that this is the optimal mix for long-term memory. Too many intuitive ideas or too many counterintuitive ideas result in poorer recall over time. This is true for fiction as well as religion.[53]

*   *   *

Ideas pop up, and some of them get around with amazing speed and stick around with surprising tenacity. People are curious and want to understand the world they live in, and many people turn to a belief system to get answers. I turned to science, but many turn to religion and cite getting a better understanding of the events in their lives as a chief reason. Let's face it: some things in this world are just hard to understand. They're complicated. Exacerbating this problem is the fact that certain phenomena, particularly at the subatomic or galactic levels, operate according to rules that violate common sense. These two factors make some scientific explanations, frankly, unbelievable.

It's been said that quantum mechanics is something "our brains just aren't wired to understand."[54] At the quantum level, objects disappear in one place and reappear in others, have indeterminate locations, and can pop into existence from nothing. It's hard to believe because it's just too weird, too unlike the world we live in and understand. We evolved in "middle world" (the perceivable scale of the world, larger than quantum and smaller than galactic), and we've evolved to be able to understand how medium-sized objects behave. Indeed, quantum mechanical theory is so strange that physicist Paul Davies said that quantum theories are "almost impossible for the non-scientist to discriminate between the legitimately weird and the outright crackpot."[55]

Scientists themselves are not immune to these biases either. In 1979 Carl Woese developed a new classification of life based on data from DNA. It suggested three main branches of life, which he called domains. Animals and plants were just a small branch of one of the domains. Many scientists, especially botanists and zoologists, were slow to accept Woese's classification. Woese said, "Biology, like physics before it, has moved to a level where the objects of interest and their interactions cannot be perceived through direct observation."[56] We see animals everywhere—doesn't it make sense that they would be a bigger part of the taxonomy?

Ernst Mayr criticized the new scheme, arguing that one of the reasons to reject the new scheme was "the principle of balance."[57] Balance is a pattern that Mayr found so valuable that it made him prefer one scientific theory over another.

Our love of patterns contributes to our enjoyment of a great variety of experiences. Explanations with patterns, good and bad, attract us for the same reasons. Beware!

4
INCONGRUITY

Absurdism, Mystery, and Puzzle

Incongruity often generates a desire to comprehend. In spite of what you might think of your fellow human beings on a day-to-day basis, we are the most intelligent and intellectually curious species around. The main reason we've come to dominate the planet as we have is because we've outsmarted the other species we compete with. Primates in general are curious, but humans take curiosity to the limit.

During our evolutionary history, once we occupied our niche, our intellectual capacities and our brain size exploded. Many studies have found that for humans in general, bigger brains mean smarter brains, accounting for 16 percent of the variance in intelligence among humans.[1] Theories abound for the specifics of why the brain got so big so fast, including selection pressure for social competition, sexual selection, gut reduction, hunting, and niche construction. But all agree that our brains got bigger because they made us, in some way or other, smarter.

In fact, one of the reasons childbirth is so painful and dangerous for human beings (more than for any other primate[2]) is that there has been an evolutionary tension between a head that wants to get bigger and a pelvis that wants to stay small. If women's pelvises were any bigger, they'd have trouble running.[3] As a result, the infant head can barely fit through the birth canal. Infants' brains at birth are as big as they are going to get, unless we're going to be a completely C-section species. We evolved to be able to do more and more things, and this evolution made our brains get bigger and bigger. We are bipedal probably because being bipedal gave our ancestors a better view.[4] Walking on two legs made the pelvis small, and gave us better use of our hands. This happened about 250,000 years ago. The brain also already uses a quarter of cardiac output, so there might be energy limitations at work as well.

Recently (in the past 20,000 years or so), human brains have been getting smaller, having lost as much volume as a tennis ball. We're not sure why. One theory holds that our brains have gotten more efficient. Another holds that it's part of our self-domestication: as our bodies got smaller, we required smaller brains to operate them. (I will return to the topic of human self-domestication in chapter 6.) A related theory holds that with civilization we just don't need to be as smart to stay alive. However, given that intelligence worldwide tends to rise every year, I have my doubts about this theory. In the last 200 years, however, brain sizes in humans have been growing again, probably due, mostly, to better nutrition.

To deal with the small size of the birth canal, one evolutionary strategy is that we're born with our skulls in pieces, the better to fit through the small birth canal. This is why babies have that soft spot on their heads. The bones of the skull fuse together later. The bones in the mother's pelvis shift around too. But there was still a head size competition. Evolution needed to find yet another workaround. How can brains get smarter without getting bigger?

Here is the strategy that evolution ended up using: rather than being hard-coded how to act in every situation the world might throw at us, we evolved to have a general-purpose learning system. That is, rather than being born knowing everything we need to know (as "precocious" species are), humans and other apes are relatively "altricious," born knowing very little but with an amazing ability to learn what we need to as we experience the world.

Birds, in contrast, are quite precocious. Some birds start hopping around and looking for food within seconds of hatching. Humans, in stark contrast, are the most altricious species known, and helpless for a very long time after birth. As evolutionary scholars Peter Richerson and Rob Boyd put it, "We are the largest brained, slowest developing member of the largest brained, slowest developing mammalian order."[5] We have to learn to survive and our parents have to take care of us in the meantime. Traditionally, this intense parental care lasted until puberty. Nowadays, it seems to last until the offspring are leaving graduate school at age thirty-three. Or maybe that was just me.

Paradoxically, our utter inability to take care of ourselves as babies is a key to our success as a species. Rather than being made perfect for our environments right out of the gate, we can adapt to almost anything the world presents us with. This has allowed human beings to live in an astonishing variety of natural environments, from the icy areas of the northernmost parts of North America, to the deserts of Africa, to the rain forests of South America.

Not only has this ability allowed us to thrive in many natural environments, but it allows us to live in different cultures. We adapt to the world we're born into, and part of that world was created by changes other people have made to it. This allows for a positive feedback loop: we are smart, we make changes to our environment (writing books, building cities, etc.), and our children learn that environment as normal, and eventually they make changes to it, physically and culturally, themselves.

Humans are more altricious than other species, for sure, but it might be that some individual people are more altricious than others. These differences in altriciousness might account for differences in intellectual ability. Turns out that the smart kids younger than eight years of age have an unusually *thin* cerebral cortex![6] But by the time they are in late childhood, they have a thicker than usual cerebral cortex.[7] Perhaps these kids are smarter because they are more altricious than their peers—they are more competent later because they are less competent earlier. They're more built to learn. If all this is true I would predict that gifted children are actually less competent in the world before age eight, because, being less precocious, they start with less, but learn more.

If we're not born with built-in routines of how to interact with the world, then we must learn how to do it. A ramification of this is that we have evolved a great *desire* to learn things about our environment.

As psychologist Alison Gopnik puts it, if seeing pattern is the "aha" phase, the experience of incongruity is the "hmm . . ." phase.[8] To make an analogy with eating, the desire for understanding, often caused by incongruity, is like hunger, and the experience of under-standing, which is a kind of pattern perception, is like the pleasure of actually eating.

There is an entire subfield of psychology based on the desire for new information: the "looking paradigm." It's used to find out what babies understand. They are shown impossible situations (demon-strated in puppet shows). When babies see something they think is interesting (usually something new, unusual, or impossible), they look longer at it (just as you would).

When children play with blocks, they are learning how the physi-cal world works through exploration, using basic hypothesis testing. Knocking down the pile of blocks is just as important for learning as building them up in the first place. This desire to learn never goes away. As a result, we have a desire to figure things out, to make sense of chaos, and we get a rush of pleasure when we do. When we experi-ence things that violate our expectations, we respond with increased attention and thinking about that inconsistent information. The more unexpected an event is, the stronger our emotional reactions are.

Along and in the Ottawa River, near where I live, people stack rocks on top of one another. The most interesting are the ones that seem the most improbable—large stones on top of small ones, or stones balancing with their smaller sides down. Locals are most in-trigued by the ones that make us wonder how they manage to stay up. The childhood fascination with blocks—making towers and knocking them down—manifests itself in sculpture as adults.

Beauty and interestingness are different things. We find some-thing interesting when we feel there is more to discover about it. We are curious. As we learn, and find patterns, we might find the stimulus more beautiful but less interesting. Any stimulus, as we experience it more, provokes a weaker and weaker neural response. This adaptation happens with the senses as well as with emotions. For a work to stay interesting after multiple exposures, the artist must find a way to counteract this adaptation. Great works of art are, experts say, endlessly fascinating.

Incongruity is the flip side of our desire to find patterns. Too little order is confusing, too much order is boring. The sweet spot is that area where tantalizing contradictions are visible, but the stimulus gives us an inkling of a hidden order that can be figured out. The notion that there's a hidden order draws people in. Repetition is related to complexity—an increase in repetition means a decrease in complexity. We like to look at things we can make sense of, but we also like to be challenged.[9]

We want this sweet spot not only over space, as in visual arts, but over time as well. If you experience too many complicated stimuli, you will start to find them tiring and prefer things to be more simple. When a stimulus has an incongruity, and a resolution to it (either in the stimulus itself or created in the audience's mind), it takes advantage of our love for incongruities as well as patterns. The mind is always trying to minimize surprise and confusion. It does this by seeking out incongruity and making sense of it—turning incongruities into patterns, as described in the last chapter. Overexposure to pattern such as symmetry and balance can lead to a lowering of arousal due to habituation.

There's a sweet spot.

One study by education scholar Hy Day presented people with visual images that varied in things like asymmetry, number of elements, and so on. The simpler patterns were initially viewed as more pleasing, which is to be expected. However, over time the pleasingness rapidly diminished. The moderately complex images got *more* pleasing (presumably as the participants perceived patterns in them) and then at some point started to decline in pleasingness.[10] This is what one would expect with the tension between pattern and incongruity. People see something moderately complex, and they are driven to find patterns in it. As they habituate to the patterns, they lose interest. The most complex images were the least popular, because the people could find no patterns in them at all.

Computer games, too, are the most fun when they consistently challenge the player enough, but not too much, and gradually increase the difficulty of the challenges. *Halo 3,* a popular Xbox 360 game, was specifically made so that each obstacle was tough enough to be interesting and challenging, but not so hard that people got frustrated. Microsoft's human-computer interaction laboratories were used to help make the game like that.[11] They were going for the sweet spot between ease and difficulty, understanding and confusion, pattern and incongruity.

A study by Peter Hekkert of people's taste for product designs found that people most preferred designs that looked quite conventional—with one unusual feature. The completely conventional designs, as well as the truly avant-garde designs, were preferred less.[12] In landscape art, as well, people prefer moderately complex landscapes. Impenetrable jungles and simple plains are less attractive.

However, these findings need to be tempered with some knowledge of how people tend to behave in psychology experiments in general. It is known that when asked to rank a bunch of things, they tend to prefer things in the middle of whatever range they are presented with. That is, people tend to like things in the middle, no matter where that middle happens to be. This is the "central tendency effect." This means that we might not be able to predict the relationship between the complexity of a particular thing and how interesting or pleasing it's going to be, because doing so depends so much on the context in which it is presented. Also, there is some counterevidence to the sweet spot theory. One study of aesthetics, by psychologist Flip Phillips, found that people preferred the very simple or the very complicated imagery to imagery in the middle of the complexity scale.[13] In a different study, led by psychologist Colin Martindale, people preferred more complex paintings.[14] Both of these studies serve as counterevidence to consider, but it is interesting that they had different results.

The artistry in music performance is not in the composition of the piece, but in the manner in which it is performed. The musician can choose, for example, to play in a more or less expressive manner. Psychologist Daniel Levitin created versions of a Chopin piece that varied in expressiveness. The least expressive versions played exactly on beat, and they sounded dull. But if there was too much expression, the piece became chaotic.[15]

Even in fantasy and science fiction, good storytellers tend to stick to the relatively, but not overly, familiar. Filmmaker James Cameron reflected this aesthetic when describing his film *Avatar:*

> If you're outlandish all the time, you've got no place to hang your hat. People have to feel connections to things that they recognize, even down to the design of the Na'vi. There's no plausible justification— unless you go to some really arcane explanation—for the Na'vi to look that human. It's just that science fiction is not made for a galactic audience. It's made by human beings for human beings.[16]

Indeed, even science fiction needs to be a bit conservative about the future. Famed science fiction writer William Gibson said,

> If one had gone to talk to a publisher in 1977 with a scenario for a science-fiction novel that was in effect the scenario for the year 2007, nobody would buy anything like it. It's too complex, with too many huge sci-fi tropes: global warming; the lethal, sexually transmitted immune-system disease; the United States, attacked by crazy terrorists, invading the wrong country. Any one of these would have been more than adequate for a science-fiction novel. But if you suggested doing them all as an imaginary future, they'd not only show you the door, they'd probably call security.[17]

In film, there is a style of editing called continuity editing that is designed to make the film easily comprehensible. One of its rules is called "match on action." It involves placing a cut where a character is doing some physical action, such as sitting down. Seeing the beginning of the action in one cut and the action finished in the other helps make the cuts appear more continuous.

Sergei Eisenstein, an early Soviet filmmaker, did not like continuity editing, because he thought audiences should have to do some intellectual work to understand a film.[18] Sometimes we're in the mood for a book, or perhaps a film, that is challenging and interesting, and other times we just want pop culture that requires no effort on our part. The sweet spot moves depending on mood.

Personality traits probably affect our taste in incongruity as well. One of the "big five" personality traits—those that social psychologists believe generally explain personality—is called "openness to experience." In a study by psychologists Gergory Feist and Tara Brady, people who scored high on this trait preferred art works with more weirdness, dissonance, and incongruity.[19] I expect that factors such as mental exhaustion and stress contribute to what kind of complexity we want. One finding by psychologist Piotr Winkielman that supports this notion is that when people are feeling insecure or sad, they prefer familiar images.[20] It takes more mental work to process challenging stimuli. Sometimes you're in the mood for Bruckheimer, sometimes you're in the mood for Eisenstein.[21]

The relationship between order and incongruity is a fascinating tension. Personal *discoveries* of order and meaning are more compelling than order that is obvious from the outset. It has been suggested that the discovery of an object or pattern is more pleasing when it takes some effort. To discover order means that it is not obvious at first. This means that there is an initial impression of either neutrality or incongruity. The discovery makes the stimulus feel deep and meaningful. The perceiver feels proud to have found hidden depths.

As sweet as this sweet spot is, people (usually) do not skip to the last chapter of a mystery novel. Rather, we like to put ourselves in states of curiosity with the expectation that the curiosity will be satisfied. Although people will often choose to have everything explained, studies show that people are happier, and their pleasure lasts longer, if there is still some uncertainty left over. Sometimes the pleasure of the resolution is too tempting, even though the mystery might ultimately make us happier. This is the pleasure paradox.

Anthropologist Pascal Boyer's theory of supernatural agents being minimally counterintuitive ideas is another expression of this sweet spot theory. Intuitive ideas are easy on the mind and benefit from facility of processing, but they are not particularly memorable or noteworthy. One bit of counterintuitiveness makes the idea of the supernatural being just mentally challenging enough to capture our interest.

The exploration of uncertainty and the resolution is something that can unfold over time. This is obvious with music and narrative, but it happens with painting and sculpture too. One cannot focus on every aspect of a painting all at once. As we experience a work of art, we see new patterns. Great works of art afford new insights with repeated exposure and study. The art stays the same; we're the ones who change.

Some arts are immediately pleasing to anyone, but for many kinds of art we have to earn appreciation—as familiarity grows, we learn more patterns. As we have seen above, connoisseurs' ability to see more patterns makes simpler works more boring to them, but they have a greater ability to appreciate complex works. They can see patterns that laypeople cannot.

Turning from art to activities, the concept of flow, pioneered by Mihály Csíkszentmihályi, is a feeling one gets while engaging in certain activities, characterized by absorption in the activity, forgetting the self, and positive emotions. Some get it surfing, some drawing, some bartending. What makes his theory relevant to the sweet spot I'm describing is that getting to it requires a bit of challenge but not too much.

Returning to computer gambling addictions, discussed in chapter 2, people have described using the gambling machines in a way frighteningly similar to flow. One addict says "It's like being in the eye of a storm, is how I'd describe it. Your vision is clear on the machine in front of you but the whole world is spinning around you, and you can't really hear anything. You aren't really there—you're with the machine and that's all you're with."[22]

Recall that we can view high dopamine levels as causing a person's sensitivity for pattern detection. People with too much dopamine, such as schizophrenics, find meaning in meaningless stimuli. Explanations, works of art, and ideas in general can vary in how incoherent they are. It stands to reason that people with more dopamine will have their sweet spot pushed more toward the nonsensical than people with less. This theory predicts that people with high dopamine will prefer absurd art (e.g., surrealism, absurdist theatre, ghazals) more than people with normal dopamine levels, and people with low dopamine will prefer (relative to others) art with more literal, obvious meaning (e.g., landscapes, portraiture, most novels). Similarly with ideas, low dopamine should predict a preference for clear explanation, and high dopamine, in contrast, should prefer obtuse explanation. I call this the "dopamine incongruity hypothesis." It has yet to be tested.

*   *   *

Play can be understood as a response to incongruity. It is a fuzzy concept in English, describing everything from a child playing house to tournament poker games to hockey games. I look at play as any kind of interactive entertainment. Passive entertainment, such as watching television, is a kind of noninteractive play. One essential element of play is that it involves dealing with some kind of make-believe world. That world might be cards and the rules of poker, or a baseball game, or a make-believe space station kids imagine themselves to be in.

Human beings are not the only animals who play. Birds and other mammals play, and learning seems to be its evolutionary function. Predators play by stalking, wrestling with, and pouncing on each other. Hunted animals play by leaping and running around. Human play can be looked at in the same framework. Sports prepare us for physical feats we might need for survival or mate attraction. More intellectual games such as chess and poker stretch our mental abilities.

At the time of writing, 90 percent of children in the Western world play computer games. Although computer games are often thought of as a waste of time, studies show that people who play computer games can respond faster to things (for example, identifying whether a bunch of letters is a word or not) without sacrificing accuracy and are more creative. It's even good for their vision. Players of computer role-playing games are better at planning and strategic thinking. Some neuroscientists believe these games teach people to learn in general. Surprisingly, the best kind of game for learning seems to be the violent first-person shooters. Playing violent military games such as *Medal of Honor* is better for you than playing puzzle games like *Tetris* or word games, as measured by a variety of tests of visual abilities. Violent military games can even increase reasoning about scientific material.[23]

Arts of all kinds inhabit the sweet spot between pattern and incongruity. Incongruity can take many forms, from formal visual aspects such as asymmetry, to more subtle mysteries in visual arts.

When we are looking at real places as well as images that depict places, real or otherwise, we have an urge to explore unknown territory. We like to see scenes that look like they would reward us with more information if we were to explore them. Interior designers, architects, and planners have hit upon this wisdom in their design of art galleries, museums, and parks. When we walk by and can see through the doorway that there is only a single room beyond, we are less likely to enter the room than if that room contains a door or a wall that might be hiding another passage. This is because we think we've seen everything there is to see and lose interest. So gallery owners often put in a wall that hides the back wall and part of a side wall. This makes people curious to see what is on the other side of the wall. Likewise, in museums, dead-end rooms are more likely to get a mere glance from the doorway. This is called "exploratory movement." This is probably why modern museums are set up as a collection of connected rooms, rather than as rooms that branch off of a central hallway, like many office buildings and homes are. Park design takes advantage of this too. A trail that goes straight is less interesting than one that winds around. If you can't see what's ahead, you want to explore. Similarly, residential districts, where beauty is more important than transportation efficiency, are made of winding roads. They invite exploration and are more attractive.

These examples are all from designed real-world experiences. But the same effects occur with static paintings and sculpture. Paintings that feature hidden places to explore are particularly compelling.

Our desire to figure things out attracts us to contradictions and impossible objects depicted in art. The work of M. C. Escher is perhaps the clearest example of this. Many of his illustrations explicitly depict impossible scenes. Just as babies spend more time looking at impossible situations in laboratory situations, we love to gaze at Escher's prints.

Some paintings are easy on the eyes. They're just plain beautiful. Others are difficult. But there are three different *ways* that paintings can be difficult: they can be perceptually complex, horrific, or surreal.

Some paintings are just complex, visually, and they take time or effort to understand. Cubism and various forms of abstraction, in addition to paintings that are quite realistic but involve a great deal of detail, fall into this class of difficult paintings. Horrific paintings have disturbing subject matter, but need not be difficult to make sense of. Like horrific films and books, they are compelling because we are attuned to attend to what we fear. Surreal paintings might not be disturbing, but they present a world unlike our own. The paintings of René Magritte are often peaceful but profoundly weird. The mind struggles to make sense of them. They are compelling for different reasons. Surreal paintings draw us in because the incongruity makes us want to solve the puzzle or contradiction they present. Certain of Salvador Dalí's paintings are both horrific and surreal, packing a compellingness one-two punch.

Incongruity can also be found at the level of color. Artists often decide to use a particular palette, the set of colors used in a work of art. Most of these colors look like they belong together, but a few others will appear as outsiders. For example, interior designers and decorators often will pick an "accent color." Such colors have been said to maintain interest in the audience.

A similar effect can be found in music.

In general, music gets much of its emotional impact from delaying and manipulating gratification based on the listener's expectations. It is seen in electronic dance music, a form of popular music including subgenres such as techno, trance, and jungle. Typically, a track will begin with a simple sound, such as a drumbeat, repeating bass line, or melody. As the track progresses, other elements are gradually introduced. Your mind is fed new things at a certain pace. As it becomes predictable, it changes.

Pop songs have a similar structure, but rather than new elements being introduced after some number of bars, the repetition occurs at the level of verses and choruses. It is common for a song to have two verses, a chorus, a verse, a chorus, a bridge, a verse, and then a few choruses. We get used to the structure of the verses, and then the chorus arrives, providing us with something fresh. As the whole verse-chorus structure begins to get familiar, a bridge arrives. The bridge fits, musically, but is often substantially different from the rest of the song. Then we are satisfied by the return to the verse-chorus structure for the end of the song, which we take pleasure in recognizing.

Repetition is important for music at the level of particular recordings. In fact, the best predictor of how much people will like a song is how many times they've heard it. Marketing researcher Mario Pandelaere found that when we hear two versions of a song, we tend to like the one we heard first—the one we are already familiar with.[24]

This appears to happen with musical genres as well. Before we are familiar with a particular genre, it all sounds the same. Usually bad. Once we have familiarity with it, we learn to appreciate the differences between the pieces. We get pleasure from recognizing bits and we might learn to like it. Like getting used to any new, unfamiliar form of art, getting used to new music genres often requires some exposure so that we can learn what things to listen for.

Familiarity might even affect what sound formats we like. In a six-year study of his students, music scholar Jonathan Berger found that every year more and more students preferred music encoded as MP3s as opposed to higher-quality sound files![25] As MP3s become more common, the particular distortion that MP3s have becomes normal, expected, and preferred.

Sound, and therefore music, is made of patterns of vibration in the air. Most music is composed of sounds that have a regular vibration. Many other sounds, such as that of a cough or a tree falling, have an irregular, chaotic pattern. Regular vibrations have a constant frequency, which is interpreted as a musical note.

With musical training, notes are heard more often—even when a more irregular pattern is played. In fact, untrained people are more sensitive to pitch discrimination than trained musicians! This surprising finding is because of the mind's tendency to autocorrect. Just as you might not notice if someone says "kitar" rather than the proper "guitar," people who are familiar with music hear a note that is close to (but not quite) middle C as middle C. This autocorrection is called "categorical perception," which is our tendency to perceive things in terms of the categories that we're used to. Interestingly, Indian music uses different notes that are closer together than those in Western music. As a result, when they hear a vibrato (that wavering sound that opera singers use when singing a single note), people more familiar with Indian music don't hear it as one note at all, but as a wavering sound that can be interpreted as agitation.[26]

Western musical notes are organized into keys, such as C minor, which are sets of notes that sound right together. Western musical composition tends to focus on the notes in the chosen key. Music that remains in one key is said to create a sense of stability and calm. Notes from outside the key can feel intrusive, but can be interesting in the same way that an accent color can be. Music that uses all the notes more or less equally (ignoring key) sounds dissonant, abrasive, and less coherent. Such music is also harder to remember. This is not to say that this kind of music can't be compelling—it can be employed to connote dark forces and atmosphere.

Many people think that music in minor keys sounds tense or sad, relative to music in major keys. This has been found to be true experimentally, in both Western and south Indian music. Even sad *speech* tends to produce notes that fall along minor scales, according to work by neuroscientist Daniel Bowling.[27] The saddest music has minor modes, low pitch, slow tempo, and dissonance. Although there appears to be some cross-cultural support for this idea, musician David Byrne reports in his book *How Music Works* that prior to the Renaissance in Europe there was no connection of minor keys and sadness, and that much Spanish music uses minor keys for happy music.[28]

* * *

Music videos are a fascinating new art form because, until Internet videos became available, they were the only way to get avant-garde film pieces to the masses. Music video directors could do crazy things given their short time scale and lack of narrative constraints. Long avant-garde films exist (*The Cremaster Cycle* films of Matthew Barney come to mind), but such works are a difficult sell to the masses. As I would predict, a study by Georgia Tech psychologist Fredda Blanchard-Fields showed that watching music videos (as compared to dramatic television shows) encourages people to think more critically and draw more conclusions about what the videos mean, presumably because of their weirdness.[29] When people see incongruity, they try to figure things out.

I believe that one of the reasons we find dance so watchable is because it violates the normal biological motion we're used to seeing. Most of the time, this means it is more beautiful than normal motion, tending toward the smooth and fluent.[30] But some dancing is deliberately awkward. Examples include butoh and popping.

Butoh is a dance form that typically involves grotesque movements, sometimes evoking derangement or sickness. Popping is a hip-hop dance form based on quickly contracting and relaxing muscles to make them jerk unnaturally. "The robot" is a popular popping dance. In both butoh and popping, the unnatural movements of the dancers evoke interest in the audience, among other ways, by violating our expectations of how bodies normally behave. Musician Brian Eno described this movement as "somadelic," like a psychedelic but for the body.[31]

This is in contrast with ballet, which, if not exactly natural looking, appears (to the Western eye, anyway) to be an idealized, beautiful style of movement. Butoh and popping are compelling because the dancers look like people with something awry; ballet dancers look like people who appear to be more graceful and beautiful than normal, an exaggeration of Western ideas of physical beauty. It is, in some sense, a caricature of beauty.

Ballet dances and Jessica Rabbit both work, in part, because of the *peak shift principle*. Peak shift is a concept that comes from animal learning. If an animal, such as a rat, is trained to react to a set of stimuli, the rat will react more strongly to an exaggerated version of the same stimuli, even if the rat has never before experienced it.

To put it in human terms, if a person has good associations with big balloons, he or she will get a great reaction from encountering huge balloons. Jessica Rabbit is like the big balloons, and not just in the obvious way. In general, this class of exaggerated experience is known as "supernormal stimuli." If men like a small waist, full lips, and wide hips, then men will find *very* attractive an artificial stimulus that features a waist-hip ratio exaggerated beyond anything they've ever seen. Like Jessica Rabbit, many statues of women from ancient times have exaggerated buttocks and breasts. Some have called them fertility statues; I prefer to call them Pleistocene pornography. The peak shift principle has been suggested to be a major force in artistic appreciation in general.

Peak shift effects can also work over evolutionary timescales. Suppose a creature lives in high grasses and the taller ones can see over the grasses to look out for predators. The others in the species might evolve to find tallness sexually attractive, because it results in taller offspring. This evolved preference would put selection pressure on them to grow *even taller,* perhaps even far beyond the height of the grasses. Sexual selection might even generate creatures so tall that the tallness is overkill in terms of its original purpose (spotting predators) but still adaptive in terms of sexual selection.

Ballet, I conjecture, looks so beautiful (to Western eyes) because it is a peak shift from what we consider graceful motion in everyday life.

Another interesting aspect of dance is that even when it incorporates movements that might be familiar, it does so in an unusual context, often without any other physical objects. Movement in dance can set our mind reeling, seeking a goal-based explanation for what we are seeing without the contextual cues to make it possible.[32]

*   *   *

Can our love for incongruities help explain why we like certain foods? Although taste in food varies from culture to culture, there are some constants, to wit: protein, fat, sweetness, and saltiness. These tastes signal the presence of nutrients that were difficult to come by in our ancestors' situations. However, there are many tastes that are acquired. Who really enjoyed his or her first taste of oysters, hot peppers, raw onions, or even the ubiquitous cup of coffee?

Some cultures delight in cultivating tastes for things that are considered disgusting by the uninitiated. The Japanese eat the puffer fish, and some diners request that enough of the neurotoxin be left in so that the lips and tongue are numbed. Every year people die from eating this. Sixteenth-century Europeans developed a taste for meat that was just this side of rotting, based on what they thought the peoples of ancient Rome and Greece ate.

Why would cultures make these choices? Science has some partial answers. Spicy foods, such as those found in Thailand, Mexico, and India, release endorphins when consumed. We literally get a pleasure boost when we feel the pain of eating spicy foods, and psychologist Brock Bastian and his team found in a study that feeling pain actually alleviates feelings of guilt.[33] There are also biological reasons to eat spicy foods—spices prevent spoilage. This is why spicy foods tend to appear in warmer climates, according to a survey by biologists Jennifer Billing and Paul Sherman.[34]

As for why people might cultivate a taste for almost-rotting meat, I don't think anyone knows for sure, but psychologist Paul Rozin has suggested that we experience "benign masochism" over the mastery of mind over body when we eat painful foods such as chili peppers.[35]

*   *   *

Just as we might get bored with a predictable song or story, we also will get bored with a predictable sporting event. Even though we might root for a particular team, and very much want that team to win, we prefer a close game to one where the team we're rooting for completely demolishes the other. Why is this? Close games are more interesting because we don't know what is going to happen. This fact suggests the counterintuitive idea that we want our favorite teams to win, but only by a little.

*   *   *

Let's use the notion of incongruity to explain popular quotations. Suppose you were to hear someone say that it's important to believe in some ideology, and it doesn't really matter which one, because if you don't, you'll be more likely to be swindled. Upon hearing this, you might reflect on whether or not this is true and perhaps ask why someone would believe it. Lacking some evidence, or reasoning, or at least some anecdote, you'd probably be unlikely to believe it or have a drive to repeat it to other people.

Compare this to hearing someone say "If you don't stand for something, you'll fall for anything," which more or less says the same thing, but sounds so much better. Why? There is pattern, in that *something* sounds a bit like *anything,* and there is also the contrast between *stand* and *fall.*

When I gave my first TEDx talk, I monitored the tweeting that people did, and the thing I said that people tweeted the most was my statement: "It takes a firm understanding of reality to make compelling fantasy." This statement has a hint of contrast, an apparent contradiction in it. People love this. You'll notice that a great many famous quotations contain ideas that have at least a superficial contradiction (e.g., "art is the lie that tells the truth").

Take the common phrase used in educational studies and minimalist art, "less is more." On the surface, of course less isn't more. It's less. What it really means is "less is better." We can all intuitively feel that "less is better" is less sticky, less riveting, than "less is more." The apparent contradiction draws us in.

Churches love to put sayings on their signs outside. I like to look at these and reflect on what makes them compelling. Often it involves incongruity or pattern: "Do not wait for the hearse to take you to church" almost rhymes, and "Forbidden fruit creates many jams" is incongruity in the form of a pun.

As we will see, incongruity is the essence of all humor.

* * *

Although most of us take laughter and humor for granted, they are mysterious things. Most contemporary theories of laughter and humor (the study of laughter goes by the amusing name of *gelotology*) involve surprise at a perception of incongruity. The incongruity theory of humor holds that we find things funny because they juxtapose ideas normally thought of in different contexts. The incongruity leaves a mystery to be figured out.

The relief theory of laughter and humor holds that laughter originally was a signal that something thought to be dangerous turns out not to be. In other words, a false alarm, or a benign violation. The sound made by our ancestors, this precursor to laughter, was contagious in the sense that an individual hearing it is likely to imitate it.

This idea, that humor occurs when there is a feeling of danger along with an assurance of safety, explains the occurrence of laughter in many things patently unfunny, such as the experience of riding a roller coaster. A friend recounted a story to me of how he and his buddy were mugged at gunpoint. After the mugger left, the friends looked at each other and burst into laughter.

The common use of profanity in stand-up comedy likely supports this idea as well. Profanity is, by definition, socially unacceptable, and is usually used by someone who is angry or hurt. Its use in stand-up comedy is effective because it involves a signal for danger, but is delivered in a safe setting so it makes us laugh. Language use requires the newer parts of our brains, but profanity uses our older brain, particularly the parts that control emotion and moving our bodies. Hearing profanity creates a powerful impact in the listener, and a study by linguist Jean-Marc Dewaele found that profane words were remembered about four times better than other words.[36] Can you believe that shit?

It would be interesting to determine if the same parts of the brain were used to perceive both humor and profanity. Many people believe that the use of profanity in stand-up is a cheap short cut and they have a point: it makes a joke evoke a laugh the same way a naked women in a photo makes you look. It's not clever, it's just a trick that nearly always works. Many comics create very funny routines without profanity. On the other hand, swear words touch us in ways that other words cannot.

Tickling, which is not funny, but generates laughter anyhow, is also explained by the false alarm theory. Tickling tends to only work on physically vulnerable parts of the body. You're being touched in a place that can easily be hurt, but it's not hurting.

Often a joke will require its audience to bring in knowledge it has about the world to get the humor, but telling the audience this information just prior to the joke kind of spoils it. The incongruity explanation endorsed here predicts this because if you tell the audience the information needed to get the joke, it reduces the surprise necessary for the joke to be funny.

The false alarm is one kind of incongruity (between danger and safety). Cognitive scientist Bruce Katz has a neural theory how this happens.[37] A joke or story has a set-up, which makes you expect a certain context or outcome. This prediction becomes active in your brain. When the incongruous information appears (e.g., the punch line), a new, unexpected context or outcome is activated. For a brief time, both the predicted and the perceived are active at the same time. According to Katz, this association results in pleasure and a perception of something being funny.

It could be that the incongruities perceived in funny situations prime the brain for detecting associations between distant concepts. Experimental participants who heard a stand-up routine performed better on word-association puzzles right afterward than those who didn't.

Jokes have become such a common phenomenon that they themselves can be the subject of jokes, such as the classic "Why did the chicken cross the road?" The answer, "To get to the other side," is funny because it's not clever, violating our expectations of what a joke should be.

*   *   *

Sometimes people like things *because* they are confusing and hard to understand. To explain this I created the concept of *idea effort justification*.

The harder you have to work for something, in general, the more you value it. This effect is called *effort justification* and is used to explain, in part, why fraternity hazing works so well.[38] The pledge appreciates fraternity membership so much in part because he had to work so hard to get it. This probably is also related to the process of getting and valuing a PhD, but that hits a little too close to home and will not be discussed further.

I have extended the theory of effort justification to the realm of ideas: the harder you have to work to come up with an idea, or to understand something, the more you will appreciate it. *Meaning is more valuable to a person if it is attained through mental effort.* This is idea effort justification. It predicts that an idea (e.g., a belief, interpretation, or meaning) that is inferred or otherwise discovered by the person through effort is valued more by that person than the same idea that is simply presented to the person. Idea effort justification happens for five reasons.[39]

First, we get a rush of pleasure when we perceive that we have figured something out. If there is no difficulty in comprehending something, the feeling of accomplishment is lessened. But that "aha" moment of understanding something difficult associates pleasure with the discovered idea.

Second, it is a reduction of cognitive dissonance, a classic finding in psychology. If you've worked hard for something, you look for (and possibly invent) a good reason you did so in order to make your life make more sense. In cases where we put in effort, or endure something difficult for little gain, we experience dissonance, which is uncomfortable. So our mind takes steps to resolve it. Perhaps by convincing ourselves that we didn't really work hard after all, or that we enjoyed the process, or that what we got out of it was well worth it. How would this work for ideas? When something is difficult to understand, and we put in some effort to get meaning out of it, our minds might try to endow that meaning with more value in order to resolve the dissonance created by the effort it took to get that meaning in the first place.

Third, we like ideas that we perceive to be ours. Psychotherapy uses a principle like this called "nondirective therapy," in which the therapist doesn't just come out and tell the client what is wrong (even though it might be perfectly clear to the therapist), but rather tries to get the client to discover it for himself or herself. The idea is that the clients are more likely to accept something they concluded themselves than something told to them explicitly by the therapist.

You might have been in a situation in which a group comes up with an idea in a meeting, but later several of the people at the meeting believe that it was *their* idea. You also might have let people think your good idea came from them. People want good ideas to be theirs.

Fourth, when you try hard, you find meaning for yourself and, when you do, I conjecture, you pick the meaning that's most precious to you—the one that resonates most deeply with what you believe: the meaning that's compelling. We tend to choose the meaning we already agree with—evidence suggests that ambiguous information gets interpreted in favorable ways.[40]

There is a fifth reason why self-generated interpretations are viewed so favorably: they are simply more easily brought to mind.

In the previous chapter, I described how ease of mental processing makes us believe and like things more. When we figure something out, not only do we remember what we've figured out, but we also remember all the reasoning and justifications that went into coming up with it. When one reads a difficult text, one will speculate on what the author had in mind and what he meant by this or that word. All these inferences are stored in memory, linked to the final interpretations. Ideas in memory that are well-connected to other ideas (in this case, justifications) are more easily remembered. The fact that information created by one's own mind is better remembered than information simply read is known as the "generation effect."[41]

When writing is clear, one does not have to work hard to understand it. But if one slaves over a difficult article, whatever meanings one manages to scry from it will seem more valuable, if only to justify for oneself why one spent so much effort trying to figure it out in the first place.

Another reason is that when we find meaning ourselves, we find those meanings that we already believe to be true. Some texts, such as postmodernist works, poetry, and many religious scriptures, are written to encourage different interpretations in different people. Different interpretations will result whenever people are in a situation in which meaning is ambiguous and they have a favorable attitude toward the author—which is most of the time. People tend to seek out information they believe they will agree with (this is the congruence bias). In these cases, they give the text the benefit of the doubt; they assume it means something correct. So they come up with interpretations they already find plausible. So of course they like the text—it's telling them what they already believe! Confirmation bias kicks in (the tendency to better remember and pay attention to information that supports what they already believe) and they find the text compelling.

If meaning is even more satisfying and rewarding when you have had to work hard to find it, then the same meaning might be less compelling if it was directly understood from a clear sentence as opposed to heavily interpreted from an obscure one.

The fact that easily remembered ideas are more likely to be believed works *against* clearly written text! When reading something
clear, one does not need to come up with justifications for it and,
as a result, it will be (all else being equal) remembered less well and
believed less.

Okay, but why would these effects occur for postmodernism and
some writers in the humanities but not for the sciences? In science,
the quality of your results is much more objective. The data, in some
sense, speak for themselves in a way they do not in the arts and
humanities. You can be a pretty lousy writer in science but still get
published if your findings are important enough. Not so for literary
criticism. If there is little or no data to speak of, on what should
someone's work be judged? Complexity of the writing steps up to the
plate as a criterion.

One might counter that scientific text is just as dense and impenetrable as anything in the humanities. One only needs to open a
biology or physics journal to see text that is nearly impossible for a
nonexpert to understand. What's the difference? Couldn't postmodern text similarly be simply a matter of jargon-ridden prose that is
intended for experts in the field?

Not so fast. Scientific text can certainly be jargon laden. Here's
an example from a biology abstract: "RPA stimulates BLM helicase
activity as well as the double Holliday junction dissolution activity
of the BLM-topoisomerase III complex."[42] I have no idea what this
means. The key difference is that for scientific writing, the goal is
that for an expert reader there will be a single interpretation. This is
not so with postmodern writing, which is written to be interpreted
in multiple ways—even by experts in the field.

Has the analysis of the works of Immanuel Kant been done to
death? A philosopher I spoke to chuckled at the idea. It seems to be
that Kant, famous for being abstruse, provides a bottomless well of
opportunity for interpretation. This is striking to me because great
works of art are described the same way—the longer experts look at
certain paintings, the more they reveal. This suggests that obscure
scholarly writing is (or is like) an art form in itself, more akin to

poetry than to scientific writing. To understand it, we need to understand how idea effort justification is part of what makes poetry compelling.

I never much liked poetry. When I came to Carleton University, there was a Monday night writer's circle. We would read our stuff aloud, think about it, and offer feedback. I was writing short stories, but most of the participants were poets. So I ended up thinking really hard about these particular poems.

A magical thing happened. I loved many of them! It was not until I put in the hard work of trying to find meaning in poetry that I finally understood what was special about it. Even now, some of my favorite poems of all time are written by Carleton undergraduates, because those are the only ones I've taken the time to find meaning in.

This idea can be tested. If poetry is more appreciated when heavily interpreted, then reading it on paper, and having lots of time with it, should produce more appreciation than hearing the same poem read aloud a single time (because it's hard to review a poem that you only hear once). I would also predict that if people like a poem read aloud, they are likely appreciating more surface-level features— word choice, rhyme, and literal meaning.

You can find a parallel to this effect in the artificial intelligence (AI) of art, a field that tries to make computer programs that create art. A general finding in the field is that the more work the audience has to put into the artwork to appreciate it, the easier it is for an AI to make stuff that people think is good.[43] For example, computer programs can make some interesting haiku, but are as yet incapable of making engaging novels, or indeed, novels that even make any sense at all.

Poetry can be thought of as the opposite of clear writing. Some poetic traditions explicitly and deliberately obfuscate language. Studies show that readers of the same piece of poetry can differ widely in the interpretations offered, and often they base them on associations between words and personal experiences.[44] However, unlike analytic philosophy, that's the whole point. We appreciate poetry in part because of idea effort justification.

What is striking is how much appreciation of poetry is like interpretation of religious texts and myth.

In many religions, as we grow up we listen to sermons and engage in rituals over and over for years. As Harvey Whitehouse says,
"the knowledge that one has endured for years the burden of routin-
ized activity and strict discipline elicits a marked reluctance to 'give
up' lightly or to tolerate the waywardness of others."[45] We have a
drive to continue to believe in something to justify the time and effort we've already put into it—that's effort justification.[46]

But *idea* effort justification can explain some of why we find
religions compelling, too.

Some have argued that religious texts should be interpreted metaphorically, rather than literally. As mythologist Joseph Campbell
puts it, it's a mistake to read myths (religious or otherwise) as prose
rather than as poetry.[47]

Metaphors can carry enormous meaning. In literature, Macbeth
describes life as a "brief candle." Metaphors allow us to communicate a great deal with just a few words.

A similar view can be taken with religious scripture, too. There
is a literal meaning and a metaphorical one, and for some the truth
is one and for some the other. Should heaven be interpreted as an
actual location or a state of being that can be reached during our
lifetime? Interpreting text literally, as fundamentalists try to do, requires *less* reading-into than a metaphorical approach.[48] As such,
idea effort justification should have a stronger affect on metaphorical readings than literal ones.

Recall that people remember self-generated ideas better. When a
person perceives an idea, that idea is associated with other things in
memory and the perceptual context in which it's perceived. These associations allow retrieval of the memory in the future. For example,
if a green dog tells you ice cream is poisonous, you might be better
able to recall the idea that ice cream is poisonous when you think
of a green dog. The more links one has to an idea, the easier it is to
recall, because there are more ways for one to retrieve it. One way

to get lots of links to an idea is through elaboration. Thinking about the idea, how it relates to your life and other ideas, etc., creates many links. It is probable that when generating ideas on one's own, there are more links to the idea than if the idea were simply presented.

Self-generated ideas are better remembered. But why should this increase belief or value in those ideas? For this, we return to the availability heuristic, which I discussed in chapter 3, on patterns: the more easily an idea is brought to memory, the more probable and common it is assumed to be. Thus, an idea that is better remembered is (unconsciously) perceived to be more probably true.

Religions can benefit from incongruity in ways other than through idea effort justification. Philosopher David Hume, in the chapter "Of Miracles" in *Philosophical Essays Concerning Human Understanding,* writes that we have a tendency to believe in miracles because of their surprising nature. Incredible things generate agreeable feelings of surprise and wonder, and the association of those feelings with the miraculous explanation makes us more likely to believe it. It's the surprising event, the incongruity, that makes us look for some supernatural meaning.

If a creed's belief system contains statements that are falsifiable, there's a chance that they'll get, well, falsified. For example, suppose a religion maintains that there are gods living on the clouds. I mean literally on the clouds. It follows from this belief that if you flew up there you'd find them in a visible, physical form. But once people actually start flying in planes and fail to find those gods, that belief, and by association the religion, loses credibility.

New religions are being created all the time. Perhaps two or three religions are created every day.[49] A tiny number of religious ideas stick around, getting passed on to others, while most others fade away. And what do we find whenever we have diversity, heritability, and differential reproduction? We get evolution by selection. The evolution of species is only the most famous example of it.

Not only do ideas evolve by being passed from one person to another, but they evolve (that is, slowly change over time) in people's individual minds. Different parts of your mind and brain are generating ideas, and some are remembered, some are forgotten, some are compelling. These ideas change and compete with other ideas. There is a marketplace of ideas inside your own head. Sometimes you like an idea enough to repeat it to someone else.

We can look at religions that are successful and begin to understand what properties make them so. From this perspective, it is clear that if a religion maintains beliefs that cannot *in principle* be found to be false, then they won't ever be disproven. To return to our example, a version of the same religion that holds that the gods in the clouds are made of some immaterial substance that cannot be directly perceived will have a better chance of surviving natural human skepticism (and eventual exploration of the sky) than the original, literal form of the belief. A better version would also hold that the statement is metaphorical—*well, not really in the clouds*. The moral is this: religious beliefs *benefit* from being undisprovable, as philosopher Daniel Dennett points out.[50] That is, unfalsifiable and metaphorical religions have a survival advantage over those that are not. Indeed, much religious text is metaphorical, allowing multiple meanings for statements such as "God is my shepherd." Evidence from the psychology laboratory of Himanshu Mishra suggests that vague information gets interpreted in a favorable way.[51] If you interpret something favorably, you're more likely to believe it—because you already believed it to begin with.

From a scientific perspective, falsifiability is a *good* thing for a theory. If a theory is falsifiable it means that there is some kind of evidence that could, in principle, be found that would show that the theory is false. When one tries and fails to find this falsifying evidence, it shows support for the theory.

Because science is so good at testing theories and because technology keeps giving us better and better ways to observe our world, it seems likely that as technology develops, the religious beliefs that we cannot ever find to be false will have an increasing survival advantage. That is, the more science and technology we have, the more important it is for a religion to have beliefs that are unfalsifiable. I call this the "increasing religious unfalsifiability hypothesis": over time, religions will, more and more, have beliefs that are unfalsifiable.

Imagine that someone is starving to death on a raft on the ocean. She prays for help, perhaps to a god or perhaps to the spirits of her ancestors. Which response seems more likely? (A) the supernatural agent makes a hamburger, root beer, and French fries appear, or (B) the supernatural agent makes a fish jump from the water into the boat.

Even an atheist will agree that A sounds kind of silly and B sounds more likely. I will coin the term *minor miracle* for a supernatural event that could be perceived by a nonbeliever to be a natural event. In contrast, a *major miracle* is one that clearly seems to be breaking the natural order of things. The hamburger and fries appearing out of thin air on a boat in the middle of an ocean would be a major miracle.

Most people's *personal* experiences with miracles are of the minor variety. Many people *believe* in major miracles, but they know of them, for the most part, second hand: either they read about them in scripture or in a book or heard about them. This makes them much less credible, because of people's tendency to create tall tales.

Let's look at the phenomenon of UFO sightings. It's easy for someone to say they saw a light in the sky that they didn't understand. It's a better story to say they saw a spaceship. With nobody to say they didn't see it, why not make the story a little better? There's some evidence that people do this—the rate of reported UFO sightings has gone *down* since the widespread placement of cameras in phones.[52] Why? If these UFOs were actually spaceships, we would have seen a steady reporting of UFO sightings, but an *increase* in photographic evidence. But that's not what happened. People who report seeing a flying saucer can now legitimately be asked, "Well, why didn't you snap a picture of it with your phone?" My interpretation of this is that most historical UFO sightings were embellished.

Let's get back to miracles. Many people with the same intuition about the likelihood of minor and major miracles also believe in an omnipotent god. That is, a god who can do *anything*. Why would a supernatural agent disguise boons and curses in terms of natural-looking events? Put another way, why would an omnipotent god choose to help someone with a minor miracle rather than a major one? A religion will have trouble surviving with too many claims of major miracles. The consequence of this, in terms of which religions last and which fade away, is that religions who claim that miracles are minor have a better chance of lasting.

Jesse Bering recounts the behavior of a church group that interprets natural disasters as signs that God disapproves of our society's increasing acceptance for gays:

> Members of the Topeka, Kansas-based Westboro Baptist Church, a faith community notorious for its antigay rhetoric and religious extremism (they run a charming little website called GodHatesFags), see signs of God's homophobic wrath in just about every catastrophe known to man. To them, the natural world is constantly chattering and abuzz with antigay slogans.[53]

If an omnipotent god wanted us to know of his disapproval unambiguously, then wouldn't he use a major miracle to communicate it?[54] This is where some religions latch onto the idea of *faith* as a virtue—that is, belief without reason or evidence. Because there is no indisputable evidence of the existence of souls, spirits, or gods, the "evidence" has to be *interpretation* of minor miracles. It is sometimes said that God uses minor miracles to *test our faith*. Religions have evolved, over time, to make some kind of sense in a world that would look much the same if they were *not* true.

Religious miracles and magical belief systems describe a world that would look to the casual observer to be nonmagical—it allows people to simultaneously believe scientific as well as biological causes. That is, people can understand that a virus causes a particular sickness, or that termites caused a house to collapse, but at the same time seek an explanation for why it happened *to this particular person at this particular time*. In many cases, science's answer is that there isn't any "reason" at all. This can be unsatisfying, and religion enters to fill the void.

Another powerful way for an idea to be unfalsifiable is for it to be simply incomprehensible. Take, for example, the Holy Trinity of Christianity, a doctrine that holds that three persons are one being. This strange idea has received a good deal of theological attention, in part, I would imagine, because it contradicts common sense. But even if we are to make some kind of sense of it, it's hard to imagine how science could test it.

Religious people deal with such evidence in a different way—because interpretation of religion is often metaphorical, disconfirming evidence triggers reinterpretation, not rejection. The religion is *assumed* to be true, so other beliefs have to make way so that the religious beliefs can remain so.

Confusing statements are compelling for other reasons: there is an inherent beauty in mystery and any meaning you get out of it is hard won and therefore more valuable to you. The term *mystery* is even a technical term in the Catholic Church for something divine that cannot be explained. In the religious context, I believe that what is working so well is the generation of the feeling of awe, which is triggered for experiences with two features: vastness and an inability to fit the experience within the existing structures we have in the mind. In some male initiation rituals, paradox is built right in. Anthropologist Michael Houseman found that in one ritual, boys are instructed to wash in mud puddles. They are beaten if they don't for not doing as told, and beaten if they do because they get dirty.[55]

Clarity of explanation is also compelling, for another reason, and religious clergy can be pulled in two directions—cultivating mystery on the one hand and clarifying themes on the other.[56] But in general, analytical thinking is the enemy of religion. In support of this, a study by psychologists Will Gervais and Ara Norenzayan found that people who are more analytical tend to be less religious, and even making people think more analytically makes them less religious.[57]

*   *   *

Pascal Boyer has a fascinating theory of why public religious rituals work the way they do.[58] Some rituals mark a change in social status—becoming an adult, being married, graduating from a university, and so on. However, many of the events that the rituals are marking happen, in reality, rather gradually. The function of social ritual is to specify a precise moment, even though that moment might be somewhat arbitrary, for society to treat these changes as having occurred. This is why it's important that coming of age and wedding rituals are public.

So what does this have to do with religion? It turns out that there is no clear line to be drawn between religious rituals and nonreligious rituals. People can find them meaningful with or without the existence of supernatural agents and some rituals involve supernatural agents only peripherally.

Because people tend to treat the subjects differently after these rituals, it seems as though the ritual actually *caused* the change, rather than just marked a time to acknowledge the change. After the wedding, everything was different. But how can this be? How can a bar mitzvah *actually* turn a boy into a man? That's the incongruity. This need for explanation is where gods and spirits come into the picture to resolve it. Cultures create an explanation involving supernatural agents to understand what might otherwise appear to be a common-sense-defying event.

Intuition provides answers, not explanations. Religion fills in the blanks.

*   *   *

Riveting incongruities have three types: absurdity, mystery, and puzzle.

Magic shows are a great example of absurdity at work. Once I saw a live Penn and Teller performance. For most of the show I was amazed at what was happening on the stage. I, like most of the audience, had no idea how they did what they were doing. Like babies watching impossible puppet shows, we reveled in the incongruities on the stage in front of us. At one point, however, they performed a complicated trick that took over five minutes to execute. It was impressive and I don't *know* how they did it, but I have a pretty good idea: at one point, Teller picked up a piece of paper that was supposedly in a sealed glass container a moment before. It would not have been difficult to use basic sleight of hand to switch that paper with another. I believe the whole five-minute trick hinged on this one sleight of hand that took less than two seconds. So what appeared to be huge and complicated was actually done by the same sleight-of-hand techniques that your uncle uses when he pulls a coin out of your ear. What's important about this example is that *the solution to the incongruity is much less interesting than the incongruity itself.* Although conspiracy theorists might not want to believe it, sometimes big, complicated events have simple explanations. We find both magicians and self-proclaimed psychics fun to watch—the difference between them is that the magician doesn't try to convince the audience that he or she is actually capable of using real magic. The magician might talk that way, but it is a convention that is understood by the audience. While our old brains are fascinated with the incongruities we see, our new brains know that it's all a delightful trick.

If magicians are like fantasy fiction writers, then psychics are like those who write "nonfiction" describing fictional events.

Psychics try to entertain your whole brain, old and new, by making you really believe it. They use some general guidelines to make the tricks more believable. For example, tricks like telekinesis must be small: if you levitate yourself and do flips, the audience will assume you're using wires, but if you concentrate and move a paperclip a few inches, it strikes people as more believable. So-called psychics also make their shows more compelling by telling the audience that they have merely better harnessed a power that lies latent in all of us—appealing to our sense of hope. Since there is no psychic ability (as many, many years of research has shown), they are frauds who appeal to compellingness at the expense of real understanding. Some might actually believe they are psychic, doing tricks such as "cold reading" without knowing it. They deserve less blame, perhaps, but their actions are just as destructive, in that they perpetuate belief in the paranormal.

Incongruous phenomena like psychic shows intrigue us because they make our minds search for meaning. This mindset we are put into makes learning more likely to happen.[59] One is less of a passive observer when engaging with the absurd.

Likewise, many fantasy books, particularly those written for children, such as *Charlie and the Chocolate Factory,* the Oz books, and the Alice in Wonderland books, feature absurdity of a kind novelist Anthony G. Francis terms "inexplicable wonderfulness." The reader of these stories does not believe that the author has an internally consistent model of the narrative's world that could be figured out. As a result, the inexplicable wonderfulness is appreciated aesthetically and not for the possible solution to the incongruities that might be found with enough thought and discussion.

Many stories from the horror genre make use of absurdity with a feature I call "inexplicable awfulness." For example, in the Clive Barker novel *The Damnation Game,* a character appears for a short scene to escort the protagonist. She has no lips. This creepy image is never explained. Nor does the reader feel the need for an explanation—perhaps the mystery is intrinsically riveting and makes the image more frightening.

There is a danger, though, that if the incongruity is too great, the audience will not have enough hope that it can be resolved. That is, if the audience feels there is no hidden pattern, no solution to be found, the lack will reduce the compellingness of the narrative for some audiences. Even in the no-lips example above, as intriguing as it is, I would predict that if the image had some information that could be interpreted as a clue to understanding, it would be even more compelling.

Other things are compelling for the presentation of the incongruity followed by a resolution, such as mystery stories, many jokes, and scientific explanations. In a mystery novel, the revealed solution, in concert with the proposed mystery, is more interesting than the mystery alone. I will refer to this class of things, the resolved incongruities, as *mysteries*. This clever tactic takes advantage of our love for incongruities (in the beginning of the narrative) and our love for patterns and figuring things out (at the end). We are delighted twice.

Scientific communication works as a mystery, and the resolution to the scientific puzzles are interesting. When the resolution fails to be compelling, it leaves the audience feeling that the theory has drained the beauty from the incongruous phenomena, which had its own absurdist beauty. Is the knowledge that the hormone oxytocin causes attachment an insult to the bond someone feels with his or her spouse of forty years? It unweaves the rainbow, as Keats said.

The final class in the taxonomy consists of incongruities that can be figured out but are not presented explicitly. I call these resolvable incongruities simply *puzzles*.

Riddles are set up as puzzles. A riddler sets one up and invites the audience to figure out the answer. If the audience correctly figures it out, it is a success, and the riddle is appreciated as a puzzle. In case of failure, the audience gives up, and the riddler then reveals the solution. In this case the riddle is appreciated as a mystery. Unlike riddles, jokes are essentially mysteries and the polite listener will not guess the punch line.

Because of idea effort justification, audiences feel a swell of pride and happiness at finding the solution to a puzzle all by themselves. It's been found that people understand characters' mental states in literature even better when those mental states are not made explicit.[60] This suggests that puzzles have the potential to be the most rewarding kind of incongruity of all. With puzzles, the audience gets to appreciate so many things: the initial incongruity, the pleasure of knowing the solution, the pride of having discovered it themselves, and an increased value of the found solution due to idea effort justification.

The *Star Wars* universe is an example of a set of artworks that work together to generate puzzles for the audience. It is one of the most fleshed-out fictional worlds ever created.[61] It has *canonicity,* a term coined by T. S. Blakeney to refer to an imagined world that, throughout the various stories and related art works, maintains an internal consistency. Online discussion boards for *Star Wars* feature endless discussions, with participants trying to figure out, for example, why some Jedi vanish when they die and others don't.

My beloved enjoyed a puzzle experience while fulfilling her prenuptial agreement to watch all six of the *Star Wars* films with me (not in the same night; I have a heart). At one point while watching the prequels, it dawned on her that the Clone Wars were orchestrated by the same man who was leading both sides. Nobody in the films ever comes out and says this, but it's something a viewer figures out. For her, it was a jaw-dropping realization. When something dawns on you, that's the satisfaction of the puzzle.

All riveting incongruities attract us because of a bit of piqued curiosity. We see that something needs resolution and we are drawn to it with the subtle promise of a solution we might discover or that might be revealed. The word *mystery* should be used when the solution is difficult to arrive at, and interesting in its own right. *Puzzle* should always be used if the author is confident that the audience will be able to figure it out, because idea effort justification will make that solution better than if it was handed to them, as it is in a mystery. *Absurdity* should be used to encourage meaning-making in the audience, or when you have some reason to believe that the audience members are secure, bored, or have high dopamine levels.

To reflect briefly on the very book you're reading, it is mostly mystery: I present problems to pique interest and then present solutions I hope the reader will find elegant and satisfying. It is also peppered with absurdism—there are still things in the world that are misunderstood.

5

OUR BIOLOGICAL NATURE

The physical nature of our senses puts severe constraints on what we find compelling. There's an obvious regularity in visual art: it is created to manipulate color within the range of light that human beings can actually perceive. There are some works of art that use ultraviolet light: they are conceptual pieces, because we can't perceive light in that part of the spectrum. We also don't make art that consists of manipulations of magnetic fields, because we can't perceive them. If we'd evolved in a world without a moon, the earth's magnetic field would be much stronger and more animals might have evolved to detect it, according to physicist John Barrow.[1] If this had been the case, we might (if we came to be at all) have whole art forms based on magnetic fields. Sight, followed by hearing, is our most important sense, explaining why we have so few art forms based on touch and smell.

The receptivity of our senses, including what chemicals our taste buds react to, the frequencies our ears are attuned to, and so on, are those parts of the world that were important for the survival of our ancestors. We're unaware of the rest, except through measurement instruments we design to translate them into something we *can* sense, such as when we see "infrared" photographs.

What we commonly refer to as our sense of touch is actually a suite of biological sensory systems that perceive things such as pain, pressure, temperature, and body position. The receptors are not only in the skin, but in muscles, bones, and even the cardiovascular system.

Our sense of where our body parts are is known as *proprioception,* and this sense might contribute to our appreciation of dance. When people observe others doing physical motion, it activates the motor parts of their brains. Visual experience gets translated into proprioceptive simulation and appreciation, even if they are just sitting and watching.[2] Personally, I have often felt, while watching a dance performance, my muscles subtly, involuntarily twitch in sympathetic response to the dancers' movements.

Humans also seem to have "mirror neurons," which respond both when doing an action and witnessing an action—but only when that action is viewed as intentional. For example, the motion of falling after being hit with a board is an unintentional motion that presumably would not activate mirror neurons. Dance is full of intentional movements on the one hand, as the dancers are making decisions about what muscles to move, but dance is an interesting case because the movements often are not obviously in the service of some clear practical goal, as the motions involved with stacking wood would be. Nevertheless, neuroscientists have shown that mirror neurons are activated in humans when experiencing both music and dance.[3]

Although watching a dance performance might activate the proprioceptive as well as the visual senses, it does not take advantage of proprioception directly. However, some activities are designed to make a person experience something by moving their bodies in particular ways. I'm talking about participatory dance, such as line dancing, which is primarily meant to be enjoyed as an activity and not as a show.

To appreciate a painting, you use your visual sense. To appreciate a piece of music, you use your auditory sense. Similarly, participatory dance is actually executed to be appreciated, stimulating the proprioceptive sense. Whether or not a piece of square dancing choreography is a work of art or not is not particularly important. The point I am making is that there are parallels in how they must be appreciated.

One of the ways that participatory dances are limited is that they generally are not intending to evoke any kind of emotional range. Square, contra, and couples dancing might evoke feelings of joy or titillation, but rarely are these kinds of dances meant to evoke, say, sadness. I have been swing dancing for many years and I can't imagine how I would make up a sad swing dance routine. There could be proprioceptive "arts" that do, however.

For example, here is a simple one:

> Title: Exhaustion
> 1. Get on the floor, on your hands and knees.
> 2. Hang your head.
> 3. Slowly move your torso up and down.

This idea is that performing these actions makes the audience (in this case, both the performers and the audience) get a sense of what it means to feel exhausted. Indeed, certain body movements are associated with certain kinds of feelings. For example, bending your arm is associated with acceptance and joy and extending your arm with rejection.[4]

Although dance choreography also consists of plans for how to move, I would not consider it a prototypical example of proprioceptive art because the intended audience typically watches other people execute the choreography. If it were a proprioceptive art, the "audience" would be the dancers themselves. Of course, I would conjecture that one of the reasons people enjoy dancing in front of audiences at all is to enjoy how it feels, and in fact the act of dancing appears to have benefits for general well-being. Performing dancers probably get an artful reaction from performance. It emotionally moves them at the same time they are trying to emotionally move the viewing audience.

The group Improv Everywhere performs ingenious performances in public, often without the public knowing, at first anyway, that a planned performance is happening. One of their works involved having multiple people listen to the same track on their portable music players. The track had instructions (e.g., "jump now"). The people listening had a great time, following instructions and doing the same things at roughly the same times.

We can now see a complex of artlike entities at several levels. The choreography for a dance piece is a work of art. Like a play script, it needs to be performed to be appreciated.[5] At the next level we have proprioceptive art, consisting of the experience of the dancer executing the choreography. Finally, we have the performance as viewed by an outside observer as a further work of art. These three levels have analogs to theater, at the level of the script, the acting, and the show.

Why would it be so emotionally moving to position and move the body in certain ways?

In the past twenty or so years, cognitive scientists have found that a great deal of our higher-level, abstract concepts have a connection to more primitive, bodily concepts. This relationship is theorized to be metaphorical by linguist George Lakoff and philosopher Mark Johnson.[6] That is, we understand an abstract concept by reasoning with simple bodily experiences, such as perception or action. For example, we understand time in terms of space. In English, we conceptualize the future as being in front of us (perhaps because as we walk forward, time changes too) and the past being behind.

In contrast, there are some languages that conceive of the past as being in front and the future behind us (perhaps because we can see the past, but not the future). Their speakers' gestures reflect this too. What's amazing is that even though cultures differ in where people perceive the direction of the future, every culture thinks of the future as having *some* direction, which is, when you think about it, completely metaphorical, since, strictly speaking, time does not have a spatial dimension.

This is all possible because of our minds' incredible abilities of association. One study by psychologist Sascha Topolinski even found that people had positive associations with dialing phone numbers that spelled out happy words (e.g., *5683* spells *love*) as opposed to numbers that spelled negative words. This happened even when the letters were not written on the number pad![7]

This theory has inspired a great deal of empirical work that supports it, and papers finding metaphorical effects are constantly being published. Just to give an example, loneliness feels cold.[8] That is, people who are feeling lonely report that a room is colder than non-lonely people do in a room of the same temperature. (Rather than review all the studies here, I will mention them as needed to explain phenomena.)

We know that these associations are more than just the conventions of language, because the same sensory terms are used for the same kinds of things across languages (e.g., *hot* referring to high-arousal emotional states).[9]

We think of the up direction as being good and the down direction as being bad. This is reflected in the phrases we use: "I'm feeling up today," "Her depression was in a downward spiral," "Things are looking up." Why might this be? An evolutionary explanation is that when people are standing upright, they are more likely to be awake, healthy, and, most importantly, alive. When they are bent over or down on the ground they are more likely to be sleeping, sick, or dead. It could be that our primitive understanding of such things influences all of our conceptions of value.

One can see how this insight applies to staging in theater and choreography. Bend someone over and, all else being equal, it looks more depressing than if he or she were standing upright. This is what inspired the proprioceptive artwork I designed above. We can clearly recognize emotion in body posture and dance. Proprioceptive appreciation of bodily position and movement is effective because our higher-level concepts are metaphorically based on them.

These metaphors affect religious beliefs as well. Many world religions have concepts of some kind of heavenly afterlife. Fewer have a version of hell. Often, this "good" afterlife or otherworldly place is associated with the sky in some way. The association of "up" with goodness and divinity, as well as the association of "down" with evil has persisted across cultures and time, even for the nonreligious. This probably seems obvious, but its implications are not. Some effects of this are downright weird. A study by psychologist Lawrence Sanna even found that people were more likely to give to charity after riding up an escalator than after riding down one.[10]

There are known reasons for this. First, as described earlier, we have a general association of goodness with aboveness and the upward direction. In addition to the normal bodily activities that predict this association (we're down when sleeping or sick), scholar Jon Tolaas points out that as infants, we literally get love and resources from above: our much larger parents.[11]

Second, the upper field of vision tends to be associated with abstract thinking, religious thought, hallucinations in general, and distance (because more distant things tend to be higher in our field of view than closer things). Hallucinations more often happen in the upper field of view and eyes tend to deviate upward during them. Think, for example, of the eyes rolling back into the head during some kind of trance. Likewise, feelings of embarrassment and shame are cross-culturally expressed with the downward positioning of the head and gaze, while success and pride is often characterized by an upward posture.[12]

More broadly, there is a "social cognitive chain of being" that places gods on top, humans in the middle, and animals at the bottom. Morality is also associated with this metaphorical, vertical chain in our minds. According to this theory, beings can be thought of as being more or less morally virtuous by pushing them up or down the chain. We find animals as being more worthy of consideration if we humanize them and other people as being morally repugnant if we dehumanize them (a kind of reverse anthropomorphization). We get the reverse effect when we sanctify people as Catholics do with saints. A study by psychologists Kurt Gray and Daniel Wegner found that thinking of gods as more humanlike (anthropomorphizing them) results in people thinking of them as less morally virtuous.[13]

The consequence of this association is that people will find explanatory structures (such as a set of religious views) more compelling if they conform to the "up is good" metaphor. A religion in which a good god is associated with being below everything will seem less plausible. It's almost a law of religion—good gods have got to be above us.

* * *

The sense of smell is often ignored in artistic works, and it's under-standable. For one thing, it's a relatively weak sense. Day by day, the average human being doesn't get very much information from smell, at least compared to vision and audition.[14] Indeed, the accommodations people need to make when they are hearing or visually impaired are often obvious, but people who cannot smell anything rarely encounter difficulties. Another reason we don't have a whole lot of olfactory art is that smells are hard to reproduce and difficult to clear out once released. Imagine the difficulty of a movie theater that had an olfactory "smell track." You'd have to pump chemicals in that have the smells you want and then be able to get them out of the air so that they would not interfere with the next smell. However, even this was tried. Smell-O-Vision injected smells into the seats, triggered by the soundtrack, and Odorama consisted of scratch-and-sniff cards given to audience members. Unlike vision or audition, which carry information as a single form of energy (light and pressure waves, respectively), smell is receptive to numerous chemicals, each of which would need to be collected or manufactured. Making a music file is easy. Smell-O-Vision is a technical challenge.

Of course, there are huge industries that try to make compelling odors. Perfumes and colognes are obvious examples, but most of what we experience as taste is actually smell. The taste buds on our tongue taste only five things: sweetness, saltiness, bitterness, sourness, and umami (savoriness). A flavorist is a professional smell maker for foods. We often don't know why people respond to smells the way they do, but some smells have real but inexplicable effects. For example, a study by consumer researcher Kyoungmi Lee found that the scent of vanilla has a calming effect. Some women, inexplicably, have an affinity for the smell of gasoline.[15]

Chefs are culinary artists. Although taste is what we normally associate with what they are trying to manipulate, food is actually a multisensory art form, involving not only taste and smell, but also touch (the texture of food is important to the experience) and vision

(people won't eat food like meat or vegetables when artificially colored blue).

So smell actually gets used in multiple art forms, including perfume creation, the culinary arts, and your occasional Smell-O-Vision.

Future technologies might make the smellable arts more feasible.
For example, if we are able to stimulate the brain or olfactory nerves
directly, we will be able to send in smells as neural signals rather
than by pumping chemicals into the air that the audience inhales. If
this sounds baffling I highly recommend Ray Kurzweil's *The Singularity Is Near* for an idea of how sensory experience via direct brain
stimulation might be possible.[16]

However it is possible that you cannot make a masterpiece with
smell, as you can with painting and literature. Perhaps this is because
it's hard to put pieces of smell together in meaningful ways, as you
can bits of color in a painting, words in a poem, notes in a song, or
events in a story. You could expose someone to fifty smells, in some
particular order, but at least in the cultures we have today, we would
have a very hard time making sense of them. Even perfume shops
keep cups of coffee beans to sniff between samples, because customers' noses get overstimulated so easily and customers need to clear
the olfactory palette.

*   *   *

Vision is our dominant sense, and hearing is the runner-up. What
we like to hear is constrained by the biology of our ears. In terms
of volume, for example, there is a level of quietness that we cannot
hear because of the structure of our tympanum.[17] Rather than being
completely free to make any kind of sounds, musicians are Caged,
forced to only make sounds that we can hear, using notes we can
distinguish. Humans can distinguish one-twelfth of an octave in music (an octave is the distance between a note and the same note at a
higher pitch). The only animals with better discrimination are bats.
Macaques, for example, can only tell the difference between musical
notes half an octave apart!

Phonemes are the basic units of sound in a language. Although
they are typically thought to be meaningless, there is evidence that

phonemes actually have meanings somewhat like words do. This theory is called *phonosymbolism,* or *phonosemantics.* Margaret Magus found that, for example, the sound *gl* is more likely than chance to be at the beginning of words the meaning of which have something to do with reflected or refracted light (e.g., *glare, glint, glimmer, glow, glance, glisten, glitter*). Phonemes also carry emotional connotations, as well as shape connotations. For example, *m* sounds are associated with curvy shapes, while plosives such as the *k* sound, are spikier. In one experiment by neuroscientists V. S. Ramachandran and E. M. Hubbard, people were asked to match two shapes to two different sounds (*kiki* and *bouba*). People consistently paired *kiki* to the sharp shape and *bouba* to the rounded, amorphous shape. Associations have even been found with tastes—*oo* is sweet and *a* (as in *cat*) is sour. There appear to be similarities with these meanings across languages. Participants in one experiment could guess the meanings of antonyms (e.g., *fast/slow*) in languages they didn't know better than chance might predict. Although we can guess why these patterns are found (e.g., *sn* words are associated with nasal meanings because we use the nasal area to make the sound), more research has to be done before we will really know.[18]

It's been shown that words with front-of-the-mouth vowels (like the *ih* sound in *tick*) are preferred for small, sharp things, and back of the mouth vowels (*a* in *tall*) are preferred for dull, large things. This could be because vowels in the front of the mouth tend to be of higher pitch, and when smaller things collide with the ground or other objects, the sound they make tends to be higher in pitch, forming the association that gets transferred to word preference. Another theory is that this effect occurs because vowel sounds have different sizes in the mouth. The vowel sounds in the following words correspond to increasing size: *beet, été, pet, pat, tall.* It turns out that your tongue moves farther back in your mouth as you make each of these sounds. The theory is that perhaps the larger space in your mouth gives people an association of the sounds with larger and larger sizes as the tongue goes back and the mouth opens wider. Also, for some reason, the "smaller" vowels are associated with brightness.[19]

*   *   *

A related finding is that we prefer designs with rounded shapes, rather than pointed ones. This could be because in the real world sharp objects can harm us. Indeed, one study by medical researchers Moshe Bar and Maital Neta showed that looking at sharp-edged forms activated neurons in the amygdala, the hub of fear. Even preliterate nonsynesthetic toddlers associated the letters *x* and *z* with black, and *o* and *i* with white. It seems that jagged shapes go with black and curvy shapes with white, as there was no correlation with the *sounds* of these letters.[20] Perhaps this is because sharp things are more dangerous, and we associate black with danger.

Whatever the reason, the implication for compellingness theory is that people will find new, made-up words more compelling if their sound associations match the properties of the thing described. This has ramifications for product naming—you probably would not want to name a new knife a *shooboo*. Perhaps the name of a drug will influence whether or not people will take it. Chemotherapy drug names have significantly more sounds in them that are associated with lightness, smallness, and fastness, such as voiceless consonants, as in the drug name ThioTEPA.[21]

Another implication concerns what sometimes happens when there is a meaning that everybody understands which does not yet have a word associated with it. For example, English has the word *orphan* for a child whose parents are dead, but we have no word for a parent whose child is dead. Even though we all understand the concept, there is a "lexical gap." This gap might be filled someday if this concept needs to be communicated frequently. What sounds will fill the gap will probably make sense given the associations we already have with the phonemes in English.

There are also tendencies to interpret cadence in particular ways. For example, the way mothers tend to speak to babies (known as *motherese*) emphasizes intention. They use exaggerated speech patterns characterized by high pitches, large pitch ranges, pauses, and a slow speed. Falling pitches soothe distressed babies. Approval is communicated with steep rising and falling pitches and disapproval is spoken in a staccato, low voice. This is not merely cultural. Babies can understand approval and prohibition phrases in other languages, as was found in a study of American and Japanese babies.[22] A universal interpretation of cadence has ramifications for the interpretation of music.

*   *   *

The quality of speaker systems and sound-generation systems continues to improve, but the nature of the actual design of sound is already very good. There's an inside joke in film sound design called the Wilhelm scream. First introduced in the 1951 film *Distant Drums,* the scream has been used over and over again in different films, and unless it's been pointed out to you, you'd probably never notice that it's exactly the same sound, over and over. We're just not as sensitive to sound as we are to visual stimuli.

Yet, human beings can only detect a small range of color frequencies. The rainbow is a tiny slice of the electromagnetic spectrum. Our eyes have three working kinds of color-detecting cells, called cones. Our eyes reduce color to essentially three parameters: the amount of activation in the long, medium, and short cones. Everything else is thrown away undetected. There are many colors out there we can't see, with information about the world we can't access directly.

Trying to picture the colors we can't see, such as x-rays, is as difficult as imagining a fourth spatial dimension. And by that I mean it's impossible. We can imagine the *concept* of ultraviolet, but we can't *imagine* it vividly.

Imagine having red-green color blindness (where only two of the cones are functioning) and then suddenly being able to use all three cones. It would be an awesome revelation. The addition of a single cone adds not one color, but *multiplies* the number of colors that can be experienced. Most birds have four cones, as do some fish and spiders. And apparently there are some women (called *tetrachromats*) who have four different kinds of cones! One negative aspect of this is that artificial visual media, such as photographs, television, and computer displays, are not optimized for the few tetrachromats and will exhibit crazy variation along those extra dimensions, making them unrealistic looking. The mantis shrimp, in one of its two very bizarre eyes, has eleven cones. It's humbling to try to imagine the rich color experience of this creature.

We usually get visual stimulus from light in the environment, but sometimes we can experience light without any actual light at all. This can happen when using certain drugs or sometimes by pressing on our eyelids. The geometric shapes commonly seen in these conditions are called *phosphenes* and have neurophysiological bases. Some cultures interpret these shapes has having a mystical meaning.[23]

Just as we cannot appreciate music at frequencies beyond our hearing range, we cannot visually appreciate things too large or too small for us to comprehend. As art scholar Dennis Dutton points out, without the aid of a microscope we cannot even see the different parts of a flea in order to have any chance of appreciating its composition.[24]

* * *

Unlike literature, storytelling, and other art forms that rely on words, visual art can have a more universal appeal. Even people of cultures that do not have pictures can recognize objects in pictures. In this way, many of the visual aspects of film and theater can be appreciated cross-culturally. You can hand an excellent untranslated Chinese novel to an American child and she might not get anything out of it, but if you sit her in front of a Chinese film she will understand a great deal of what it is about, even if she understands none of the spoken words.

This is because much of human movement and expression is cross-cultural. We all understand fighting, walking, resisting, and so on. Similarly, the six "basic" emotions—namely fear, joy, surprise, disgust, anger, and sadness—have been found to be universally recognizable in facial expressions. These facts give visual media, such as film, theater, painting, and photography universal appeal when depicting human beings, especially human faces.

This is not to say that all emotions are easily read on the face. Nor are all expressions completely universal. For example, a Japanese person will nod her head and giggle to express confusion. Japanese art depicts anger with crossed eyes.

*   *   *

Some people have supposed that humans are the only creatures that can appreciate music. However, research by psychologist Adena Schachner found that species that can engage in vocal mimicry (e.g., parrots) can move in sync with music, suggesting that our rhythmic abilities grew out of our ability to speak.[25] Interestingly, Charles Darwin suggested that our ability to speak grew out of what he called "rudimentary song."

Cognitive scientist Tom Fritz exposed the Mafa (an ethnic group in Cameroon who had never heard western music) to excerpts of classical piano music. They consistently identified the same songs as sad or happy, suggesting that the emotional response to music does not depend on cultural background. Music researcher Roberto Bresin did research in which experts and laypeople moved sliders to adjust music to make it maximally sad, happy, etc. All agreed on which tempos were best for which moods.[26]

To some extent music mimics our voices. Neuroscientist Daniel Bowling found that when people (English and Tamil speakers) were speaking sadly their voices were more monotone. Similarly, Western classical music and Indian ragas both show the same pattern: sadder music has smaller intervals (the distances between notes in a melody) and happier music has larger intervals.[27]

Ultimately, much of music appreciation is metaphorical—musical elements deeply remind us of primal things we care about. Conductors have used the wrinkling of their brows to affect the sound of their orchestra. Journalist Justin Davidson reported that conductor Lorin Maazel produced a rich, honeyed sound from the string section in his orchestra by contracting his brow and a lyrical and light tone by lifting his eyebrows.[28] It makes evolutionary sense to imagine that the upward direction is associated with lightness and the downward with heaviness. Heavy things are difficult to get off the ground. What's interesting is that we describe music as being heavy at all. Lowering your pitch at the end of an utterance is associated with finality, rest, and completion. Ending a sentence with rising indicates incompletion. These speech associations have clear correlates in music.

Physically, sound is the interpretation of pressure waves in the ear, and heaviness does not enter into it. Why is a high-pitched sound interpreted as being, well, high? There's nothing intrinsically high about it. Pitches are high or low because of a high or low wave frequency, of course, but we have no direct experience of this, and further, there's nothing intrinsically high about a high number either. Why should we say that ten is higher than eight? Numbers are ordered, but there's nothing intrinsically, spatially directional about them. We perceive higher pitches to be happy and lower pitches to be sad, demonstrating the conceptual metaphors that quantity means highness and that highness means goodness.[29]

*   *   *

What we see is linked to how we move, which is a part of why we like to watch sports. One of the reasons mammals play is because play simulates practice in dangerous situations. I conjecture that sports are an extension of this practice-play (note that in English and many other languages we "play" sports). Sports appeal to our sense of physical competition; we play sports for the same reason we wrestle with our children.

We don't just play sports, though, we often just watch them. Why would we want to watch someone else playing, indeed, sometimes paying thousands of dollars to do so? In addition to the explanations provided elsewhere in this book, I believe that we vicariously experience the motion of the athletes just as we activate our motor brain areas and mirror neurons when we watch dance. So in a sense, we like to watch sports for the same reasons we like to play them.

In our minds, we *are* playing them.

* * *

Imagining yourself slam-dunking is all very well, but what of religious experiences? Scientists have found that they involve a few parts of the brain that are more and less active, depending on the kind of religious experience.

Three percent of Americans have had what are described as "near-death experiences."[30] The typical symptoms include experiences of approaching a bright light, seeing souls of the dead, feeling that one's soul has left the body, and moving to another reality full of love and bliss, or sometimes terror. These characteristics are common in theological traditions worldwide, and there are physiological events that seem to trigger these experiences. For example, many of these experiences occur to people who are not near death at all, such as those experiencing sleep paralysis, high stress, or oxygen deprivation (which causes increases in dopamine). One diabetic patient had the classic symptoms of a near-death experience while having low blood sugar. Some experiences seem to be triggered by a failure to integrate information from the senses properly, which interrupts normal representation of the self and its relationship to the world.

There are interesting brain theories about the perception of a tunnel with light at the end. Pilots under high G-forces can sometimes experience a narrowing of the visual field that is often interpreted as a tunnel. This effect can also be caused by glaucoma. Blood and oxygen depletion in the retina can cause an experience of light as well as fear.

Visions of monsters, fairies, and dead people can result from a variety of conditions, including Alzheimer's disease, Parkinson's disease, and certain brain lesions. Stimulating one part of the brain (the angular gyrus) can result in a sense of a felt presence. Degeneration in the eye, as in Charles-Bonnet syndrome, can cause visions of ghosts or fairytale characters. One theory of hallucinations is that they are the result of one part of the brain interpreting noise from another, damaged part of the brain.

Sensed presences (in the absence of an actual other person) are often triggered by barren landscapes, monotony, darkness, cold, isolation, dehydration, hunger, fear, fatigue, and sleep deprivation.[31] Of course, people experience supernatural beings without being under these conditions. According to anthropologist Scott Atran there is no known report by an anthropologist of two people perceiving the same supernatural being in the same way at the same time,[32] further suggesting that these experiences are caused by mental processes and not by something in the real world. If a supernatural being actually appeared, it stands to reason that more than one person might experience it at the same time.

The bliss felt during some of these experiences could be caused by the surge we often get of dopamine and other opioids when under severe stress. Clearly, there is no simple explanation for why people have near-death experiences.

At the beginning of this chapter I focused mostly on traits of a healthy brain, but specific brain problems have implications for what's compelling as well. In chapter 3 I talked about how schizophrenia-spectrum effects can create and support religious belief. A similar effect happens through obsessive-compulsive spectrum disorders.

Obsessive-compulsive disorder (OCD) is characterized by intrusive thoughts and a compulsion to engage in ritualized behavior to reduce anxiety. In particular, the intrusive thoughts often have to do with sexual or aggressive impulses, an aversion to or affection for certain numbers, or the belief that inanimate objects have souls. The rituals, which can absorb hours of the victim's day, often have to do with cleaning (e.g., incessant hand or wall washing), hoarding (e.g., not being able to ever throw away a piece of paper), entering or leaving spaces (e.g., having to count to a certain number before entering a doorway), and checking (e.g., returning home again and again to make sure the stove is off). Often, the sufferers themselves know, intellectually, that what they are doing is irrational but are unable to stop.[33] Failure to perform the rituals results in fear and anguish. It's the fourth most common mental disorder, affecting 2 percent of all Americans.

OCD, like many mental illnesses, is probably the result of over-activity of mental processes that normally help us. It's good for our health to keep clean. We sometimes forget things, so it's good to check once in a while. We do sometimes have to do "rituals" to prevent undetectable hazards without knowing the actual connection between the ritual and the real-world effect.

It is likely that religious rituals are compelling because they activate these same functions—the same parts of our mind that, when in overdrive, result in OCD. Indeed, hyperreligiosity is a major feature of OCD. And society's reactions to people with OCD can take a religious turn. Western society used to think that people with obsessive blasphemous or sexual thoughts were possessed by the devil.[34]

Further evidence that religious ritual is connected with OCD comes from the fact that religious rituals often have the same content and themes as OCD rituals. Orthodox religions are replete with food and body cleansing, repetition of mantras, numerology, and rules on how to enter and leave places. In Hinduism, for example, an orthodox Brahmin might spend six hours a day in cleansing rituals. Brahmins have rituals regarding which foot to put down first when getting out of bed, what they can't look at when defecating, how to enter and leave a temple, how many times they must wash each hand, how to eat, and the reciting of sequences of magic numbers. There are hours of ritual before the first meal of the day.[35] Orthodox Jews have a spectacular number of laws about food preparation and eating, ritualistic cleansing (including putting silverware in dirt), entering and leaving holy places, special numbers, the number of bones believed to be in the body, and the number of days in a year. Islam details rules about what to eat, how to enter and leave mosques, how to wash out the mouth, and which hand to wash in which sequence. When cleansing, if you touch your penis you have to start over.[36] This requirement—that if the ritual is not done correctly, it must be started over—has a striking similarity with the rituals of obsessive-compulsive disorder—the repetition is why victims of OCD often waste hours every day.

Cultural and religious rituals are compelling often because it can feel dangerous not to perform them. Just as with OCD, the failure to perform can cause a vague sense of dread, even though the participants don't have any clear idea of any mechanism that connects the ritual to events in the future.[37] Because there is no apparent connection between rituals and events, it allows for rituals featuring what appear to be meaningless and arbitrary actions to become widespread.

People with OCD-spectrum disorders are often attracted to religion. A study by psychologist Craig Gonsalvez found that Catholics have higher levels of obsessive-compulsive symptoms than Protestants or the nonreligious, reflecting Catholicism's ritualistic nature.[38] Orthodox Christianity has its share of OCD-like characteristics: the number of prayers to the Virgin Mary, details of baptism (a cleansing), magic numbers involving the Holy Trinity, etc. These are just a few examples of how religious practices are similar to OCD ritualistic behavior. In the modern Western world, people have some freedom to choose a religion or church. If people have proclivities for ritual they will be attracted to a church that offers it.

Sometimes an individual will become obsessed with religious ritual, and religious people have taken note of such obsessions. Five centuries ago Ignatius Loyola described the condition of scrupulosity as being anxious and too obsessed with doing everything perfectly, and even described how religious leaders should help people with this problem.

The OCD cleaning symptoms are examples of evolved traits that work well in moderation but become problems when indulged in in an extreme form. People have an urge to clean; people with OCD can take it too far. Why do people have this urge to clean? They probably have an evolved psychological fear of contagion. People don't want to be around obviously sick people, in general, and people don't even want to touch the things sick people have touched. People claim to have an aversion to washed clothing worn by murderers or even by people who lose limbs in accidents. A number of studies by Carol Nemeroff have shown how this "magical contagion" belief works:

people don't want an AIDS patient to occupy a hospital bed after they've left it, which implies a belief in causality, which goes, weirdly, backward through time. One experimental participant said about the idea of touching a sweater worn by someone with hepatitis, "I'd feel it was contaminated in some way, not only that I could get hepatitis from it, but that it was somehow contaminated, it's just not clean. I don't really think you could get it [hepatitis] that way."[39] The description suggests that this is an old-brain bias, a deep-seated feeling that can be at odds with what we know intellectually. If you doubt this effect, try this thought experiment: imagine how you would feel putting on a newly washed shirt that Adolph Hitler had slept in. If this thought creeps you out more than the thought of wearing a washed shirt from some unremarkable person, then you are feeling the power of this effect.

Perhaps this is why we think properties like happiness and powers can rub off onto objects too. People pay lots of money for trivial items that were owned by famous people. If the belief in good contagion was as strong as that in bad contagion, there might be comparable ideas about contact between perceived lower and upper classes elevating the lower rather than polluting the higher. But there are not. There is an imbalance—the bad can contaminate the good more easily, or more effectively, than the good can improve the bad. In Hinduism, if a Brahmin and an untouchable make contact, it's the Brahmin that is harmed and not the untouchable that is helped. What's interesting about this imbalance is that it mirrors the effects of infectious disease. If a diseased person comes into contact with a healthy person, the healthy person is in danger of getting sick, but the diseased person will not get healthy as a result of the contact. One exception to this is the "virgin cure" myth, which holds that having sex with a virgin can cure some venereal infections, such as AIDS or syphilis. It doesn't work, of course, and rather than being a harmless myth, it has resulted in the rape of young women and even children.

Unlike many dangers, microbial danger is invisible. Germs are too small to see, but they can still make you sick and kill you. But we had to evolve a way to avoid microbial danger anyway, using other indicators, other signals that, throughout our evolutionary history, correlated with infection. Many of these indicators are outward signs of sickness, such as spots, coughing, being dirty, etc. The problem is that it's not obvious how these indicators are dangerous. If you know something about the physiology of sickness, you can come up with causal connections, but of course we did not evolve with this knowledge and our gut reactions still do not use them. That is, the ancient instincts to stay clear of people who look sick work relatively independently of our reasoning systems that come up with explicit theories about illness.

Once a culture decides that a particular group has some sort of contamination as a part of their essence, then the contagion reasoning system kicks in and the idea of even touching them triggers disgust. Anthropologist Pascal Boyer reports that in one culture, people would not have a blacksmith over for dinner for this very reason.[40]

When we have an intuitive reaction that we do not understand, religion often steps in to provide an explanation. So we get ideas of certain practices or objects being forbidden for *religious* reasons. Incest is a wonderful nonmicrobial example of poorly understood intuitions resulting in religious concepts: religions the world over claim that incestuous acts can cause natural disasters.

Although there are patterns to religious ritual, in that they often resemble OCD rituals in theme, there is certainly a strong component of arbitrariness. And although religious scholars sometimes generate reasons for the various parts of a ritual, for most people explanations simply don't matter. Bells and incense symbolize nothing in particular in Hindu ceremony, and most people are fine with that. Interestingly, ritual displays are common in other social vertebrates and can appear just as arbitrary. For example, there's a bird called a manakin that moonwalks to impress a potential mate. *Superstitious* rituals also appear to be arbitrary, but I don't know of a connection except that the existence of both behaviors shows that people generally have no problem engaging with rituals that make no sense to them.

Another reason religion can be compelling is because some of its tenets correspond well with psychological experiences we, as humans, can have, such as getting high on hallucinogens such as "magic mushrooms." Psilocybin, the active hallucinogenic ingredient found in select mushrooms, has been found to cause feelings of peace, intense happiness, and a sense of the unity of all things.[41] The human mind can experience certain phenomena that religion can then interpret in a compelling way. Scientists have found there are ways to actually induce religious experience through brain stimulation. Though controversial in science, the Koren helmet is a device reported to induce feelings of a benign presence in some people. Some of these people attribute this to the experience of God.[42] Strange experiences like this get interpreted according to whatever background belief system people have. The interpretation becomes part of the remembered experience, in that we don't even see it as an interpretation at all, but as a direct perception. We then remember the experience as further evidence of the belief system that helped generate it. In this way belief systems are self-justifying.

*   *   *

One kind of religious experience involves the feeling of being out of control of one's body. These feelings are often interpreted as feeling possessed by another entity, and belief in possession is common to many religious traditions. However, the specifics of the attribution are culturally constrained, supporting the idea that possession is a confabulation generated by the one in the trance to explain the feeling. They base this explanation on ideas familiar to them from their culture, be they spirits, ghosts, deities, or demons. But if not supernatural agents, what causes the experience?

In schizophrenia, one can sometimes feel that one is not in control of one's actions. In epileptic seizures, you lose your feeling of self, similar to the feelings people report when under "spiritual possession." In somatoparaphrenia, you have symptoms such as loss of mind (e.g., possession) and loss of ownership (e.g., this leg attached to me is not mine). When people report being possessed, they report four broad kinds of entities at work in controlling them: (1) the self, (2) an alien, (3) a malfunctioning machine, and (4) an engineer. This effect can be induced with hypnotic suggestion and looked at with brain imaging techniques. These examples from mental disorders show that there is something in the mind that can cause experiences of dissociation of ownership and control. These mind functions might be the same ones in use in certain religious experiences.

Dissociative identity disorder (also known as multiple personality disorder), in which more than one distinct personality resides in a single person, has been suggested to be related to possession claims.[43] This relationship is controversial, as is the very existence of dissociative identity disorder.

Another disturbing possibility is that women who claim to be possessed might be describing a rape or some other kind of sexual assault in a culturally acceptable way. Some suggestive bits of evidence for this are (1) the primarily heterosexual male spirit and female victim relationship, (2) vaginal or rectal entry of the spirit, during possession, or exit of the spirit, during exorcism, and (3) descriptions by many of the victims that sound like rape scenarios. In cultures where being raped is shameful, a victim, consciously or otherwise, might describe the traumatic episode in terms of the supernatural.

* * *

Revelation occurs when a person endorses beliefs from what is perceived to be a nontraditional source, such as being told something by a god. There are several types, including divination (e.g., reading tarot cards), hallucination (dreams, visions, etc.), infusion of inspired wisdom, and embodiment (meeting a person believed to be a divine being incarnated).

Several religions hold that one of our primary goals should be enlightenment. There are different things that are supposed to happen upon reaching this state, and traditions disagree on what exactly the state feels like. One constant, however, seems to be achieving an escape from the discomfort intrinsic to nonenlightened existence. Some sects say enlightenment entails gaining a true understanding of the world. For some, when this happens, the enlightened person realizes that he or she is one with the universe, that the distinction between the self and the rest of the universe is illusory.

There is reason to think that this transcendental feeling is the result of quieting the left hemisphere of the brain. Neuroscientist Jill Bolte Taylor tells in her book *My Stroke of Insight* about how a stroke temporarily disabled her left hemisphere. She felt that she was one with the universe. It could be that the long hours of certain kinds of meditation required to achieve spiritual enlightenment are, at a brain level, quieting the constant chatter and theorizing of the left hemisphere.

One of the differences between the right and left hemispheres is that the right tries to capture reality more or less accurately, without embellishment or interpretation. The left, on the other hand, is constantly theorizing and finding patterns. The left hemisphere has been shown to be more important for dreams and hallucinations. We can see the curious effects of this in split-brain patients.

The left and right brain hemispheres communicate through a structure called the corpus callosum. People who have trouble with seizures sometimes get the corpus callosum cut so that seizures won't travel from one half of the brain to the other. It's akin to creating a fire line in the woods. The most remarkable thing about these split-brain patients is that for the most part their lives are unaffected. They feel no different and their performance on most tasks, following surgery, is the same.

However, because the different hemispheres control different sides of the body (the right hemisphere controls the left half of the body and vice versa), scientists can make experiments that pit the hemispheres against each other. In one experiment people were shown pictures, then asked to choose the most relevant among a group of objects. The trick was that the different halves of the brain were shown different pictures. This is possible because the right hemisphere is in charge of processing the left half of each eye's visual field, and vice versa. In one trial the left brain was shown a chicken and the right brain was shown a snowy day. Now, if the right (controlled by the left brain) hand executes the action that picks the relevant object, it picks something relevant to the chicken. If the left hand is picking the relevant object, it will pick a shovel. The weird thing in this case is that when researchers ask the participant why she picked the shovel she will say she picked it because she needed to clean out the chicken shed—an answer that has nothing to do with the wintery image that made her pick the shovel in the first place. Why? Because the left brain is, for the most part, in charge of language. The right brain has no voice. The left brain *confabulated* a reason. The real reason the participant picked the shovel was because the right brain saw a snowy day. But the left brain made up a reason to make sense of the choice, and the person, apparently, believed it. It has been said that the whole reason we have a sense of self at all is because "self" is basically a theory generated by the left brain.

Another relevant split-brain experiment concerns how the different hemispheres deal with randomness. In an experiment by neuroscientist Michael Gazzaniga, people were presented with two keys, a left one and a right one. One of the keys (say, the left) would give a reward 70 percent of the time, the other 30 percent of the time. These percentages were never explained to the participants, but through trial and error they got a rough idea of these percentages. In this situation, the most rational thing to do would be to press the left key all the time, because it yields a better return. Indeed, this is what rats learn to do in the same situation. People, however, end up pressing the left key only 70 percent of the time—some part of their brains thinks it can outsmart the system and guess correctly when the left will actually pay out, which of course it can't. This leads to less payout over time. Which side of the brain is that?

If you guessed left, you guessed right. The isolated right hemisphere will choose to press left all the time, wisely, just like a rat, but the left thinks it can outsmart the system.[44] As in most things, the left hemisphere tends to dominate when both hemispheres are being used. If you can't understand why always pressing the left button would give a higher payout, try asking your right hemisphere. Or a rat.

We might say we are of two minds about what we are. Our left brain says we are a single person, separate from the environment.

The right might see no such distinction, but it can't say anything, and we are often only conscious of its point of view when high on drugs or after thirty years of meditative practice.

Let's bring this discussion back to religion. When we have un-usual experiences, such as feeling one with the universe, they might be a result of right-brain activity. However, our religious *interpretation* of that experience is likely coming from the left brain. As you might expect, religion cannot be associated with one half of the brain or the other, and religious behavior does not preferentially correlate with activation in either one of the hemispheres.

Buddhists claim that after many years of meditation one can reach an experience interpreted as enlightenment. Meditation sounds relaxing, and perhaps for some people it is, but some, this author included, find it more like taking your brain to the gym. It's hard work. The Dalai Lama urged neuroscientists to find a way to get the same results through brain stimulation. Apparently even he finds it oner-ous to meditate for four hours a day.

Buddhists interpret enlightenment as offering a view of the world as it actually is, a special insight into the nature of existence. Spe-cifically, they see such revelations as evidence that the self, as distinct from other things, is an illusion, and that the self and the rest of the universe is a seamless whole.

It turns out that meditation reduces activation in the parietal lobe, which is the area that allows you to distinguish your self from your environment, and in the left side of the brain in general, which is associated with thinking in a more narrative and less holistic way. Indeed, people who are less in touch with their bodies (as measured by their ability to mimic the poses of mannequins) are more likely to have out-of-body experiences.[45] It is likely that long practice of cer-tain kinds of meditation allows one to willfully reduce parietal lobe or left hemisphere activation. To me, concluding that you are a seam-less part of the universe because you've managed to quiet the part of your brain that differentiates yourself from the rest of the world is like concluding that light does not exist because you've trained yourself to close your eyes.

Why might mystics see the right brain's opinion of the world as more valid? First, it's difficult to shut the left brain up long enough for the experience of the right brain to become dominant. Perhaps mystics see that point of view as valuable because it is hard to attain—as my theory of idea effort justification would predict. Second, apparently enlightenment is a wonderful feeling. In her TED talk Jill Bolte Taylor mentioned that during her stroke, it was liberating to not have the left brain chattering at her all the time. The view of reality that the right brain provides is peaceful and happy. Since people want to believe what they hope to be true, this would be another reason for people to prefer it as a clearer picture of reality.

*   *   *

In periods of anxiety, many people engage in milder versions of OCD rituals. Note how common superstitions, such as avoiding stepping on cracks and throwing salt over one's shoulder, are arbitrary actions to keep supposed bad things from happening, such as being influenced by demons or breaking one's mother's back.

I suggested above that superstitious learning happened in animals when the environment was more unpredictable. Similarly, in human beings, superstitions seem to happen more often in erratic environments. Unable to control the world, people resort to controlling their own behavior. When events are perceived as surprising or salient for any reason, it triggers causal reasoning and the mind searches for an explanation.

For example, anthropologist Bronislaw Malinowski found that Trobriand fishermen had more superstitions in the open ocean, where catches were unpredictable, than in the inner lagoon, where catches were more predictable. In baseball, batters' rituals can be extraordinarily complex. Interestingly, there are more rituals for the more unpredictable parts of baseball: hitting and pitching. Fielding, relative to hitting, is much more based on skill and less on chance, and as a result players have fewer magical rituals for it.[46]

Chaos is frightening, and when people are frightened, they are more likely to see patterns that are not there. In one experiment by

Jason Braithwaite, people were asked if they could discern any pattern in some random noise. The ones who were about to jump out of airplanes were more likely to discern something.[47] Pattern detection is not as simple as the level of dopamine—it seems that it's a right-brain hemisphere specialty. The right hemisphere picks up more pattern in random noise than the left, and paranormal believers tend to have relatively more right-hemisphere brain activation than skeptics, who tend to be left-hemisphere dominant.[48]

Religion, particularly of the kinds that have beliefs that resemble superstition, should be more common in highly chaotic and unpredictable human environments.

*   *   *

In Western culture, the belief in ghosts is amazingly pervasive. A Gallup poll in 2005 found that about a third of all people are believers.[49] Ophthalmologist William Wilmer investigated some of his patients who appeared to be living in a haunted house. They heard bells and footsteps in the night. They felt physical sensations and saw mysterious figures. They experienced headaches and fatigue. The previous owners had had similar experiences. It turned out that this old house had a faulty furnace that was releasing carbon monoxide into the house. It's odorless, but breathing it can cause visions, feelings of dread, and hallucinations.[50] When the furnace was fixed, these symptoms went away.

It might seem that the explanation wraps things up just a little too neatly. One chemical accounting for so much of ghost folklore? It makes more sense if we look at carbon monoxide not just as being the proximate explanation for an individual family's experience, but as the cause for many of the beliefs our culture holds about ghosts in the first place. That is, we believe that ghosts are more likely to be in old houses, make mysterious noises, and appear and disappear because these ideas were originally formed by people with carbon monoxide poisoning. This theory would predict that our cultural beliefs about ghost mythology got a foothold in the public consciousness around the time that carbon monoxide was prevalent in people's houses. And indeed this is true: ghost stories proliferated after the introduction of gas furnaces in homes after the mid-1800s. There are some confounding factors to this causal connection, however. For example, people of the time were fascinated with all kinds of other spiritual phenomena, including mesmerism and animal magnetism.

*   *   *

We evolved to survive and reproduce in a particular environment. There are things we need to sense, such as the sound of sticks cracking that signal the approach of another being, and things we don't need to sense, such as ultraviolet light. If we'd evolved differently, we'd find ourselves compelled by different kinds of experiences.

Small genetic differences can even affect our appreciation: some people, due to a genetic characteristic, find the taste of cilantro to be similar to soap. As such, their appreciation of cilantro-based recipes is fundamentally different from that of others.

We evolved to be careful of contagion and to protect our own, and we developed other tendencies that affect our appreciation of arts and affect the resonance we have with religious practices and explanations. Moving away from biological and evolutionary reasons for why we find certain things compelling, we now come to the mind, and how its biases affect what rivets us.

6

OUR PSYCHOLOGICAL BIASES

We evolved to survive and flourish in the world we happen to live in. But during the course of our lifetimes, we also *learn* to deal in the world we experience. Be it because of nature, nurture, or some complex relationship between the two, our minds have little tricks they use to help us get by.

Some people are generally critical of these tricks and tend to call them "biases." Others see them in a more favorable light and tend to call them "heuristics," which roughly translates as a rule that should *usually* be followed. Everyone can agree that these tricks work *best* in environments resembling that of our evolutionary history.[1] Problems can occur when these biases, evolved in one environment, are used in other environments, such the one you're in when choosing a health insurance plan.

Whether biases/heuristics are good or bad, artists and marketers use many of these tricks to win us over. I'll describe some of these biases and their effects.

* * *

How do you know how likely something is to happen? Or how common something is? Before we had statistics and systematic counting to help us, we used what we saw in our environment as a guide.

We still do. Seeing something again and again makes us think it is common and probable. We estimate how often we've seen it, in general, by how easily it is brought to memory or, in the language of cognitive psychology, how "available" it is. This is the essence of the *availability heuristic*. We base our ideas of how probable or common something is on how easily it is brought to memory. This tends to work pretty well when we are only using our sensory perception as input, that is, reasoning about what we *directly* experience. The trouble starts with communication—when we reason about what we are told, read, and see on video.

We can recall not only things we've actually seen, but also reports of what other people have seen—or their reports of other reports. As described in previous chapters, we have good evolutionary reasons to believe the stories that people tell us. Today, we have global communication and we can hear about things happening all over the world. Not only do we read about what's happening, but we also *see* what's happening with television and Internet news. Where we might have evolved some built-in filters for lie detection, video recording has been around for such a short time we can't possibly have evolved an innate lie filter for it. The result is that we, or most of our minds anyway, believe what we see and even think it's happening right before our eyes—even if it's just an Internet video. The effect of the availability heuristic combined with mass media communication is profound: we get a skewed vision of what the world is like from the news.

For example, women in their forties vastly overestimate their chances of getting breast cancer, and this overestimation correlates with news reports about breast cancer.[2] This is consistent with the availability heuristic. A study by sociologists Lisa Kort Butler and Kelly Sittner Hartshorn found that watching crime coverage in local TV news makes people more likely to think the local crime rate is increasing.[3] What most of us know of crime, pollution, and other world problems comes from news. The problem with most news is that it is in competition with other news. This provides an incentive to show compelling stories. This is taken for granted in our society, but its negative effects are underappreciated. If we are prone to find stories and statements that involve hopeful or scary things, novelty, and human drama more compelling, then the news media have an incentive to provide stories that feature these attributes, rather than the more realistic or important things that might be less compelling.

The problem is made worse because news is not only in competition with other news, but with other media that seek only to entertain. That is, you could read a novel or read the news. You could watch a movie or play a computer game instead of watching the news. If you have to present a news story that successfully competes with a team of dramatic writers, you'd better make it as engaging as you can, or you lose the ratings or newspaper sales that keep your organization in the black.

One of the biggest problems with news is evident in its very name: "news." It is an information source focused on things that have happened recently. There is a good psychological reason for this: we habituate to things and lose interest in patterns we feel we already understand. The news media did a good job of reporting the dangers of drunk driving during the 1980s. But by the 1990s people had gotten tired of hearing about it. Trying to find a new angle on traffic death, the media jumped on the idea of road rage, which is a rare phenomenon but boasts compelling features that make it newsworthy, including playing to our fears and the fact that it was a novel concept. Media dropped the problem of drunk driving. It was no longer news. It's not that the problem of drunk driving had gone away—far from it. The news just needed a new fear to monger, regardless of how important it was.[4] It deemphasizes the routine and the constant, and brings irregularities to our attention. Understanding of how things typically are is key to understanding our world. But to a great extent news tells us things that are anomalous, which we then perceive as common and probable.

Another feature of our memory makes us very easy to scare: memories are easier to recall if they are emotionally charged, sexual, or violent. Because it's easier to recall violent and frightening stories, we end up thinking they represent common or probable events.

Sex and violence sell stories, be they in the news, in fiction, or in between. The availability heuristic makes us think the portrayed events are common, which makes us think they're important, which fuels our desire to know about them, which causes a feedback loop into the media. The availability cascade recycles these ideas, making these events seem normal and the ideas the media gives us to explain them seem acceptable. Finally, the cycle peters out when people are habituated to them, whereupon the media grasps for a new terror for us to worry over.

* * *

In line with the metaphor theory mentioned in a chapter 4, it seems that certain colors have particular meanings to us. Although some have speculated that our color associations are evolved, others have shown evidence that our emotional responses to colors are based on associations we draw with things we like and dislike in our environments. But because we all live in similar environments (night is dark, etc.), we have learned cross-cultural similarities in our color associations.

The difference between lightness and darkness is the most fundamental color sensitivity human beings have. Pioneering work by anthropologist Brent Berlin and linguist Paul Kay found that cultures differ in what colors they have words for. Some cultures have only two color words, most have more. But *all cultures with color words have words for light and dark.*[5] Cultures with two color words use one word for all light colors and another for all the dark. Why should lightness and darkness have such a universal importance? Because in most real-world visual scenes, this grey-scale contrast contains the most important information.

Darkness is scary because we can't put our most powerful sense, vision, to good use outwitting predators and competitors.[6] Darkness is universally associated with fear and evil, explaining why we have phrases like *black-hearted* and why villains are usually portrayed as wearing black in fiction. Psychological research by Brian Meier supports this.[7] People associate lighter-colored pills with calm, medium-brightness pills with love and excitement, and darker pills with failure and poor health. A study by business researcher Anat Lechner found this to be true across countries.[8] In general, people prefer brighter versions of colors to darker ones. Exceptions exist, of course. In Beijing opera, a black face is good and a white one represents craftiness and cunning.[9] But these exceptions are rare.

Indeed, it isn't just black that is associated with evil and white with goodness—if you present any color in a brighter form, people have more positive associations with it. Another reason we might associate black with evil is that we are more likely to do bad things when it's dark. Research by psychologist Chen-Bo Zhong showed that even wearing sunglasses increases antisocial behavior.[10] The reason? We feel less likely to be seen.

Black seems to remind us of sin and our ancient fear of contagion and dirtiness. One experiment by psychologists Gary Sherman and Gerald Clore had people look at words presented in either black or white, and were instructed to say the word aloud. Some of the words expressed notions of immorality, such as *greed,* and some of the words were the "moral" words, such as *honesty.* People were faster at saying the bad words in black and the good words in white. This all happened too quickly to be deliberate. This effect was even stronger for people who were preoccupied with purity and pollution and for people who showed an affinity for personal cleaning products, such as hand soap and toothpaste. Interestingly, for people with an affinity for cleaning products that are nonpersonal, such as window cleaner, it was not stronger.[11]

This theory fits with the general law of contagion I alluded to in chapter 5, which holds that objects and people that come into contact with each other can transfer properties. Although there are beliefs in positive contagion, the negative contagion effect is stronger.[12] This has had unfortunate cultural ramifications. Texts associated with the American Jim Crow laws and with the apartheid system in South Africa have many references to contagion, concerned as they are with the pollution of the white essence.

It is disturbing to think that our tendency to associate darkness with immorality and filth contributes to the widespread racism against peoples of color. Perhaps knowledge of this tendency can be a step toward its elimination.

I have seen no evidence to link seasonal affective disorder (SAD) to our associations with black and darkness. Indeed, some research suggests that the common notion that weather influences happiness is not true at all.[13] It is intriguing that winter, presumably because of the relative dearth of bright light (exposure to bright light is a treatment for SAD), *seems* to generate depression in some people. Even if it does not, our associations with darkness and sadness might perpetuate the belief that it does. Even if something doesn't need explaining, because it doesn't exist, the widespread *belief* needs explaining. Another interesting angle on associating darkness with sadness is that depressed persons perceive the world as being greyer. A study by research psychiatrist Ledger Tebartz van Elst found that they are unable to distinguish the contrast levels between black and white as easily as nondepressed people.[14] They perceive a world with less color saturation. Perhaps it's not that grey days cause depression, but that depression causes days to appear grey.

Berlin and Kay found that after light and dark, red is the most common color term, cross-linguistically. That is, if a language only has three color terms, those colors will be white, black, and red (which usually covers browns and oranges). Interestingly, in child development, infants can detect darkness levels first, and then reds, before being able to distinguish other colors. Even infants prefer focal colors, such as red, to peripheral colors, such as magenta. People who lose color vision due to brain damage lose red last and recover it first. Of all the colors, red evokes the strongest emotional reactions.[15] For all these reasons, I conjecture that black, white, and red are the most powerful colors in art.

If blackness means fear and evil, and white means goodness, what does red mean? First, it seems to carry a meaning of violence, danger, dominance, and aggression. Of course blood is red. Red coloration has been associated with dominance and aggression in a number of animals (humans included) and this effect appears to be innate. Angry people are better at perceiving it. Male mandrills (the largest species of monkey) use red faces, rumps, and genitalia as status symbols, communicating fighting ability. This redness is physiologically related to testosterone and aggression levels. When males face off, the paler (less red) male usually stands down. In an experiment with humans, volunteers were shown different colored circles and were asked to indicate which would be "most likely to win a physical competition" and which circle looked "most dominant." Red won. Soccer goalies perceive red-shirted opponents to be better players, but this might be because teams who regularly wear red uniforms actually win more often! There's a similar effect of uniform color in fighting sports. It could be because it makes you play better, but could also be because seeing red makes your opponents play worse. Wearing red might make people better athletes (at least in competitive sports), but it turns out referees award more points to people for simply wearing red.[16]

Second, red appears to carry a meaning of sexuality and passion. Men find women wearing red more attractive, will sit closer to them, and ask them more intimate questions. Women also find men wearing red to be more attractive, but not more likable or agreeable—it is because they are perceived to be of higher status. Red is associated with excitement and heat[17] (longer wavelengths are perceived as warm, shorter as cool), which are a part of both sex and violence.

Seeing the color red can affect our cognitive abilities. In one experiment by psychologist Markus Maier, volunteers were asked to carry out a five-minute IQ test. They were assigned a bogus "participant number," written in either red or black, on the corner of the test paper. Volunteers whose numbers were written in red scored consistently lower on the tests. The students were also given different colored folders and then were asked to choose their preferred level of difficulty for an IQ test. Students given red folders tended to choose easier tests. Seeing a flash of red made people worse at solving anagrams. Seeing more neutral colors did not. You make more mistakes performing cognitive tasks in a red room than in a blue one. However, exposure to red during a task does not always produce worse results. Red enhances performance on detail-orientated tasks, whereas blue improves the results of creative tasks.[18] It could be that the violence association with red increases focus, which is known to be beneficial for well-formed problems and bad for creativity.

Red is used to signify wrath, passion, and violence in the arts, and Isaiah 1:18 features the line "Though your sins be as scarlet, they shall be white as snow."

If a language has four colors, it will have black, white, red, and either green or yellow. Languages with five colors will have the other, and languages with six will have those and blue.

Studies have not found associated cross-cultural emotional reactions as reliably for green, yellow, and blue as they have for black, white, and red. It could be that these meanings are dominated by culture. However, since most of the studies done are in America and Canada and still fail to find reliability, it could be that this culture does not have strong meanings for these colors independent of context. But there is evidence that the color green appears to make people more creative, and the presence of vegetation (which is mostly green) increases creativity and reduces crime.[19]

Currently in our culture we associate pink with girls and blue with boys. It wasn't always this way. We associate blue with peacefulness. In our culture, blue used to be considered a female color. One can look at the outfits of Disney princesses to see when this was true (Cinderella, Alice, Snow White). Eventually the culture came to associate girls with pink and boys with blue. There is evidence, however, that cross-culturally (Chinese and American) baby girls prefer colors more toward the red end of the spectrum and baby boys prefer the bluer ones. These tests are run on infants young enough that, scientists assume, the children could not have been affected by culture yet.

Although there have been studies that show that a certain shade of pink makes prisoners more calm, this is not a robust finding.[20]

*   *   *

A wall with a fresh coat of blue paint can be attractive, but there is little in this world more riveting than seeing an incredibly beautiful person. We tend to focus on what makes us different—she likes beards, he likes redheads—so it's easy to forget that the basics of attractiveness are cross-cultural and probably have evolutionary underpinnings.

Reproduction and survival are the two strongest forces in evolution, no matter what niche an organism occupies in an ecosystem. This means that every living thing has built-in, genetically coded properties that encourage reproduction and survival. Such strong forces affect what we find compelling.

Most human reproduction requires sexual intercourse. As such, we are terribly interested in sex. Studies by psychologist Jens Förster have shown that thinking about sex, or even being reminded of sex, makes you less creative and better at analytical problems (recall that the color red, which is associated with sex, has the same effect). Thinking about love has the opposite effect. The theory is that love makes you think more globally and sex makes you focus locally.[21]

There are many factors that influence sexual attraction, but the most oft-studied factor is physical beauty. Though there are cultural differences in what we find physically attractive, there are interesting cross-cultural trends that are predicted by evolutionary psychological ideas. Many of our preferences for physical appearance have evolutionary origins, which affect who we are attracted to, in real life, on film, and in paintings. Beauty is riveting. Even people with no culture at all—newborn infants—have a preference for what are agreed to be attractive faces, as was found in a study by psychologist Judith Langlois.[22] We can fairly accurately judge the attractiveness of people even when we see their faces for as little as 13 milliseconds. This is so fast that it doesn't even register consciously. But when people are shown faces this fast and are then asked to guess at how attractive the face was, they are surprisingly accurate. This is amazing, considering they don't even remember seeing anything! The evolutionary theory is that we will have a tendency to find attractive those people who exhibit traits that will maximize our "reproductive success." Reproductive success is having kids, who have kids, and so on. It's getting your genes successfully into future generations.

The popular conception of attractiveness is that men tend to focus on the physical beauty of women and women are attracted to the social status of men. A widely accepted idea is that women and men look for different things in mates due to biological differences in parental investment in children: women are interested in men who will stick around and can care for their offspring and men are interested in women who will give birth to healthy kids. But it's not just men who say they value women for beauty: women do too. Women claim to value other women for their beauty and men value other men for their status, as suggested by psychology studies that found that both men *and* women pay more attention to men who display social dominance and women who display physical attractiveness.[23]

This looks fairly obvious if one looks at the media created to titillate men and women: pornography versus romance novels and "chick films." Pornography tends to be very graphic and sexual. The characters lack interesting relationships. Aside from the differences of the genders of the characters, gay pornography is much more similar to straight pornography than it is to what women (be they lesbian or straight) expose themselves to in order to get turned on. Romances catered to women focus on falling in love rather than sex, for the most part.[24]

But isn't this the situation only in our culture? No. It's not just our culture in which men report being concerned with physical beauty and women report being concerned with status; a large study by psychologist Todd Shackelford found it to be generally true in over three dozen cultures.[25]

But people's old brains know what they want even if their new brains do not. Although the evolutionary story of why men and women should differ in what they are attracted to makes sense— in that studies seem to show that women and men differ in what they *report* to be important—in one experiment these differences disappeared when psychologists Paul Eastwick and Eli Finkel observed whom people are *actually* attracted to. In this study of speed dating men and women *showed no gender differences* in how they weighed physical attractiveness, personality, and earning potential in the people they said they actually wanted to meet. In that same study, there was no significant correlation between what individuals said they were attracted to and what they actually were attracted to. People will report what they believe to be plausible predictions of what they will find attractive, but, shockingly, they are inaccurate.[26] As usual, our new brains are making guesses—often wrong—about the preferences of our old brain.

One way to interpret these data is as follows: men and women find physical looks equally important; these preferences are biological; and finally men and women *report* differently because of *cultural* influences. There are two problems with this interpretation. First, the gender differences in reporting are cross-cultural. Why would we find differences in reporting in over thirteen *different* cultures? Most cultures are patriarchal, but why should a patriarchal culture be more inclined to promote the idea that attractive women are beautiful and attractive men are wealthy? It does not seem to benefit men all that much, in that in many cultures across the world there are many poor, good-looking guys who would benefit from a cultural idea that women like attractive men rather than the few rich, powerful ones. The second problem is that it assumes that culture is more likely to affect our deliberative processing than our intuitive processing. But something so ingrained in culture as this would presumably affect even our intuitive ideas of what is attractive. As far as I can tell, this dissociation between reporting and actual attraction is still unexplained.

What *do* people find attractive? In this section I will focus on physical attractiveness, because that's what most of the research addresses. There's a lot of research in this field and it's growing rapidly; I will summarize the findings to date. (I'm going to talk a lot about men and women, and when I do I'm referring to heterosexual men and women. Unfortunately, there is relatively little scientific study of gay men and even less research on the psychology of bisexuals and lesbians, a scientific deficiency I hope will be remedied in the future.)

I can be deeply engaged in conversation with someone in a coffee shop, but if a beautiful woman walks by, I can be so distracted that I stop talking for a second or two. Even though it is embarrassing, it is very hard to keep from doing. My new brain cares about the conversation. My new brain knows I'm happily married. My old brain doesn't care about any of that—*can't you see that a beautiful woman just walked by?* What's going on here? Recall the old brain/new brain distinction. No matter what a (heterosexual) man is trying to work on, if there's a beautiful woman around, his old brain will be preoccupied with figuring out how to impress her, distracting him from the task at hand.

Sexual desire is a powerful motivator, and lustful thoughts make our entire brain buzz with activity. A study by psychologist Johan Karremans showed that just talking to a woman makes men (on average) dumber, as measured on a subsequent cognitive test, particularly if the man is attracted to her. This was true whether or not the man was in a relationship. The harder they tried to make a good impression, the dumber they got, because trying to attract a woman requires significant cognitive resources. Women only suffer in the analogous situation if they particularly want to impress the man in question.[27]

So what makes someone attractive? Some of the first findings showed that average faces were more attractive than nonaverage faces. It was theorized that this was because averageness was some kind of predictor of genetic fitness. It might be that average-looking people are perceived to be healthier, although its correlation to actual health is rather weak.[28]

It has also been found that people prefer roughly symmetrical faces, but people are also pretty good at detecting who's good looking just by looking at one half of their face. The waist-to-hip ratio of women seems to predict fecundity and possibly health, but according to a study by psychologists Jason Weeden and John Sabini, there appears to be no relationship between health and attractiveness in men.[29] Perhaps men's attractiveness failing to indicate anything else about him explains why women's ratings for the attractiveness of men vary more than men's ratings for the attractiveness of women—they're not correlated with anything important. As far as I can tell, we just don't know why women are attracted to certain men's looks.

High testosterone levels make the human male face look more masculine, which correlates well with perceived and actual social dominance, and testosterone levels in men predict rates of divorce, infidelity, and violence.[30] Women tend to find masculine faces sexier. Why then isn't there an arms race for higher testosterone in men? Why hasn't testosterone just gotten higher and higher over the course of evolutionary history? It turns out that testosterone interferes with proper immune-system functioning, making it a very costly hormone to have in abundance.[31] This means that if a man has a great deal of testosterone he must be tough enough in other ways to survive having a bit of a compromised immune system. In other words, the look that high testosterone generates is an indicator of good genes.[32]

Women are interested in more than good genes. They also tend to look for men who are going to be good parents. Unfortunately, it's tough to have it both ways. Experiments by psychologist Daniel Kruger have shown that women have judged photos of men with more masculine faces as having poorer parenting skills and more aggression, and men with more feminine faces to be better with parenting skills and to be more supportive and diligent.[33] As a result, females tend to have a conditional mating strategy, by which they respond to more masculine faces and dominant men when looking for short-term relationships when fertile or when cheating (which is called "extra-pair copulation" in the scientific literature). For long-term relationships, women prefer more feminine faces (signaling good parenting skills). Three to four percent of babies born to couples involved extra-pair copulation on the part of the mother. That means that three out of every hundred fathers are cuckolded! Given other studies about women's preferences for long- and short-term partners, it could be that some women want to be impregnated by a more masculine guy but have their children raised by the more feminine guy. There is some interesting counterevidence: the theory

predicts that women looking for sperm donors would prefer masculine faces, because the sperm donor is not expected to help raise the
child. Mysteriously, this is not so.[34]

Tallness, too, is interpreted as a major signal of social dominance by both men and women. This is probably true for nonhuman
animals as well. Low-ranking individuals in many species with dominance hierarchies (such as chickens and dogs) behave similarly, that
is, to appear small and nonthreatening.[35] An experiment by psychologists Brian Meier and Sarah Dionne showed the surprising finding
that our association of the upper part of the visual field with social
power means that women find pictures of men more attractive when
they are near the top of the computer screen! The same study found
that men find pictures of women more attractive if they are at the
bottom of the screen. This suggests that men have a slight preference
for women with less power. Indeed, studies show that men prefer
women with low socioeconomic status.[36]

Men, too, are sensitive to masculine faces. They were shown to
prefer the less masculine looking guys to accompany their girlfriends
on a trip. Both men and women preferred photos of less masculine
looking men to date their daughters, presumably because parents are
looking to match their daughters with mates with long-term potential in mind. Nobody has studied which men parents would choose
for their daughters for one-night stands. I can only imagine the looks
the parents would give the experimenters in such a study.

This is not to say that there are no individual differences, even
between people in the same culture. In support of biologist William
Hamilton's selfish gene theory,[37] people tend to be attracted to other
people with whom they share physical characteristics that appear to
have nothing to do with survival or reproductive fitness. There are
mild but significant correlations (the correlation is 0.2; a perfect correlation would be 1.0) for physical-trait similarities between married
partners, such as the breadth of the nose, length of earlobe, wrist mea-
surements, the distance between the eyes, and lung volume. The length
of the middle finger correlates to a surprising degree of 0.61. That's
high. Physical similarity in general even predicts marital success.[38]

This suggests that the mate choice we strive for is, at an unconscious level, an attempt to get someone with genes most similar to ours. However, there is a limit to this striving too, because when genes are too similar, as in the case of our close relatives, there is a high chance of birth defects. The Westermarck effect is our built-in way to avoid incest: we are usually not sexually attracted to anyone we interact with between birth and age six. This explains the curious fact that people who grow up on communes together tend not to marry each other. If people are around each other when they're really young, their minds assume that they're related and prevent sexual attraction from developing.[39]

A study by psychologist Devendra Singh showed that men are attracted to a particular waist-to-hip ratio (0.70) in women. This means a narrow waist and wider hips. The average woman has a ratio of about 0.83. Why do people prefer it to be 0.70? A low waist-to-hip ratio predicts fecundity and general intelligence in women and social intelligence even more strongly. Women with a low waist-to-hip ratio tend to produce smarter children.[40]

In general, a preference for a narrow waist seems to be a cultural constant and abdominal obesity, as measured by waist size, predicts decreased estrogen, reduced fertility, and risk for diseases. We find evidence of this in the literature of India and China and in descriptions of feminine beauty in British literature from as early as the sixteenth century. Although the *ratio* preference appears to be cross-cultural, there are cultural differences in *weight* preferences for women. Poor people tend to prefer heavier women and affluent cultures tend to prefer thinner women. In fact, a fascinating study by psychologists Leif Nelson and Evan Morrison found that men even tend to prefer heavier women when they are hungry![41]

If we were baboons, our female pinup models would sport big pink butts. But human females are rare among animals in that it isn't clear by looking at them whether they are ovulating. Of course, from an evolutionary perspective it is to the males' advantage to be able to figure out when ovulation is occurring. And indeed, males can sometimes detect ovulation, albeit unconsciously. Men should find ovulating women more attractive, and we'd predict that they'd work harder to impress them. A study by psychologist Geoffrey Miller found that strippers working during their peak fertility periods made an average of $70 per hour in tips, versus $35 when menstruating and $50 in between. How is this communicated? It could be subtle differences in scent, facial structure, waist-to-hip ratio, or how the women move. On fertile days, women's voices go up in pitch, their breasts become more symmetrical, and their waist-to-hip ratio is accentuated.

Birth control pills prevent pregnancy by tricking the body into thinking it's pregnant already, and men seem to be able to tell. Being on the pill reduced lap dancers' tips as well, making $37 per hour, with no peak, versus $53 on average for lap dancers not on the pill.[42]

A person's natural body smell also contributes to attractiveness, and that could be because our smell is affected by the nature of our immune systems. Since we are subconsciously looking for complementary immune systems in a mate, any potential children resulting will have a well-balanced immune system. Some scientists say that combining differing genetic blueprints for immune systems turns out to be more effective than mixing similar ones. Couples with similar immune systems are more likely to cheat on each other. Couples with similar immune systems miscarry more often. It's been theorized that this is nature's way of cutting its losses early for an offspring that won't make it anyway. There's a famous "smelly T-shirt study," by biologist Claus Wedekind, in which men were asked to wear T-shirts for a few days. Then women were brought in to sniff the shirts and rate how much they liked the smell (they never met the men). The women tended to like the smell of the T-shirts worn by men with complementary (that is, different) immune systems. In a smelly T-shirt study of their own, men found the T-shirts of ovulating women more attractive.[43] Let's hope smelly T-shirts have introduced a new scientific paradigm that will continue to facilitate more scientific breakthroughs.

The birth control pill switches women's preferences in terms of smell: when on the pill, they like men with similar immune systems.

This is bad news for women who fall in love while on the pill. When they go off the pill to try to get pregnant, they sometimes find themselves less attracted to their husbands. The theory behind this is that impregnated women want to be around kin, who have similar immune systems.[44]

There are serious real-world ramifications of perceptions of beauty. Economists Daniel Hamermesh and Jeff Biddle found that being unattractive costs both men and women about 9 percent in their hourly earnings. We tend to like attractive people more. But there are downsides to being attractive, too. Doctors tend to think that attractive people are healthier and in less pain, leading to inappropriate diagnoses or treatments. Attractive women, it turns out, should not include their photo on a résumé or job website. It has been found that it makes no difference to men, but if a woman is making a hiring decision, the picture will hurt her chances. This is, presumably, due to the competitive nature women have with regard, specifically, to attractiveness.

We perceive attractive people as being more intelligent. However the importance of looks for perceptions of intelligence drops significantly when the person we're evaluating starts talking. Behavior is a much more important cue, particularly verbal ability.[45]

* * *

Much of this book has been about how we are the same. No matter what culture a person comes from, there are near-universal preferences for certain kinds of art, products, and ideas. However, to say that the commonalities we share account for all of what resonates with us would be a gross overstatement. Culture is a powerful force and much of the variety we see in beliefs and in art is a result of cultural forces.

Music and dance scholar Alan Lomax has an interesting theory of how culture affects native dance forms (as opposed to more recently invented dance forms, such as ballet). He holds that the typical motions done for work in a culture are reflected in the complexity of the movements made in dance. For example, cultures that have no metal and must chip away at wood with other pieces of wood tend to have dances dominated by movement of limbs and the body in one dimension, such as repetitive up and down motions. Cultures that have metal use swinging motions to do work, and their dances reflect this complexity with two-dimensional motions, etc. Lomax has his critics, however, who claim that his "choreometrics" is pseudoscience and that his evidence is spotty.[46]

Artistic folk knowledge has been supported by empirical studies that don't focus on art at all. In one experiment, people were asked which direction they would associate with a particular verb. People were remarkably consistent. Verbs in general seem to have directions associated with them, as found in a landmark study by psychologist Daniel Richardson.[47] When presented with a choice of directions (up, down, left, right) people are surprisingly consistent when asked to associate them with specific verbs. *Respect* is up, *denigration* is down, due to the "upward motion is good" metaphor I've mentioned before.

If you read books on how to direct plays you will find advice that "good" motion is left to right in the visual field. Films display this sensibility as well. In *The Wizard of Oz,* Dorothy is almost always moving left to right on the screen. In the *Matrix* series of films, every time the protagonist, Neo, is getting into a fight, he moves from the left and his enemy runs at him from another direction. The left-to-right motion influences Western views of sporting events as well. A study by psychologist Anne Maas found that sportscasters considered goals scored left to right in the visual field to be stronger, faster, and more beautiful. The same result was found, with smaller effects, for violence in fight scenes. Soccer referees call more fouls when players are moving left to right as opposed to right to left.[48] It turns out that this effect has to do with the direction of writing. In English, the subject is before the object, and on paper on the left. If you ask people in Western culture to draw a circle pushing a square, they will put the circle on the left. Arabs, who write right to left, do the opposite.[49]

When watching sports, many people believe in the "hot hand," the belief that when an athlete is doing well, he or she is "hot" and will continue to do well. We think we see streaks in good (and bad) performance. Although this might be true at times, most of the time these trends are illusory, according to a study by psychologist Amos Tversky.[50] We overperceive them. Our perception of these sequences of scoring is an example of the clustering illusion. Psychologist Gerd Gigerenzer asks why people should have the clustering illusion as well as the gambler's fallacy, which is basically the opposite prediction: that when flipping a coin, a string of heads, for example, increases the likelihood of a tail. People think that the coin is somehow "due" for a tail, or that a roulette wheel is due for a black after a string of reds.[51]

My suggestion for the reason for this difference is that that we expect living things to behave systematically and predictably and we expect "random" events in the world to reflect our ideas about what's random. Since a human being is making shots on the basketball court, we use the clustering illusion to infer a hot hand, whereas we get the opposite with the coin flip. This prediction could be tested in an experimental setting, asking people to reveal either the hot hand bias or the gambler's fallacy in how they predict the outcomes of random processes that either do or do not have a human being involved. Slot machines, though they are an example of gambling, are subject to the hot hand fallacy not the gambler's fallacy. People don't stay with a machine that's losing. My guess is that people view the machine as an agent making choices rather than as an inanimate object like a die or a deck of cards.

*   *   *

I've talked about how people are all similar because of how our species has evolved and how groups of people can be different because of their cultures. People differ in what they find riveting at an individual level too, not only because of people's individual histories but also because of personalities.

The stereotypes we associate with who likes what kind of music tend to be pretty accurate. People seem to prefer art that matches their personalities. People who like rock are more likely to be artistic and anxious and not very conservative or friendly. People who like classical tend to be friendly, emotionally stable, and conscientious.

In a study of people between the ages of 17 and 40 psychologists Peter Rentfrow and Samuel Gosling found that people are pretty good at guessing personality traits from what kind of music they like.[52]

Extroverts tend to like horror and stimulation that includes sensational elements, such as action and bright colors. They tend to be sensation seekers and are drawn to media with elements like miracles and war. Most people are drawn to pleasant stimuli, but sensation seekers even seek unpleasant novelty. They also tend to like people, so they're attracted to concerts and music with vocals.

A study by psychologists Tomas Chamorro-Premuzic and Adrian Furnham found that people who are more intellectual, have higher IQs, and are more open to new experiences tend to appreciate music in a more cerebral way, focusing on complexity, and that they listen with a critical ear. The more neurotic and introverted and less conscientious use music more for emotional regulation, like for a pick-me-up after a hard day.[53]

* * *

As people get more training and familiarity with the arts, they tend to focus on more formal aspects of art—things such as spatial composition—to determine preferences. Untrained people tend to prefer realistic scenes and scenes depicting subject matter that they like.[54]

In chapter 1, I noted that paintings and photographs tended to feature human beings. In this chapter, I have shown why people might prefer to look at *attractive* people. We want to see images of people we find attractive because a part of our minds thinks they are really there and have a chance to maximize our reproductive fitness. It is kind of sad to think of a teenage boy's old brain reacting as though having a poster of a swimsuit model on his wall actually is giving him a chance to reproduce with her, but I believe that this is exactly what is going on.

This idea makes some interesting and sometimes conflicting predictions. As with everything in psychology, these generalizations will not be true for *all* individuals. People have the option to surround themselves with images of men or women, attractive or not, and of high social status, roughly equal status, or low social status. What do people actually do?

We can make a prediction based on the relative social status hypothesis I mentioned earlier. On the one hand, observers might be happy to see lower status people, as they remind them of their own relatively high social status. On the other hand, unless observers have a close personal relationship with the observed lower-status people (e.g., their children), they won't have a lot of interest in attending to them and therefore will not want to see their pictures on a daily basis. The hypothesis predicts that people will feel compelled to attend to people of *slightly* lower status. However, the rivalry can create anxiety, so observers will probably not choose to surround themselves with images of lower-status people, for the same reason that even if we enjoy horror movies, we don't decorate our houses with horrific imagery. I know several hard-core horror fans and none would put a big picture of a monster eating someone's head in their living room. When people observe others of equal status, the hypothesis predicts that they will only have images of friends. Rivals will not be on the walls because of the anxiety produced by seeing them. The hypothesis predicts that observers will not want to observe those with slightly higher status, because seeing them will cause aggression. However, the hypothesis predicts that if the observed are of much higher status, and are perceived as heroes or role models, seeing them could be inspiring. Many girls only have pictures of models and other women on their walls. I conjecture that this is akin to boys having pictures of their sports stars on their walls—they are role models to aspire to.

What do the data show? There are no data, to my knowledge, so for now I can only rely on my anecdotal observations. When people decorate their walls with images of people (those that are not explained by sexual attraction), they put up primarily photos and paintings of unknown persons, family members, or famous people. I realize that covers about everybody, so allow me to explain each one.

I predict that nobody has pictures of people they don't know unless they find them sexually attractive or find the picture of some aesthetic value—in contrast with a picture of a loved one, which need not have any aesthetic value. An example of a picture of an unknown person might be a photograph of a nude model, taken in black and white, or a pretty painting of some unidentifiable person standing in a beach scene. We don't seem to have nonartistic, nonerotic photos (akin to snapshots) of people we don't recognize.

We care about and love family members, and seeing their pictures gives us joy. More specifically, we will only display images of those family members whom we care about and love. Some of these will be of lower status, such as children, and some of higher, such as parents (until a certain point, at which parents are reduced to a lower social status, at least in Western culture).

Although it is less common for adults (in Western culture) to display paintings and photos of famous people on their walls, teenagers do it often. The depicted people are either sexually attractive, displayed in an aesthetically pleasing way, or are role models, such as famous sports figures, actresses, etc.

Ultimately, we choose to surround ourselves with pictures of people for the same basic reason we surround ourselves with any pictures: because they make us feel good in some way. They make us feel affection (e.g., pictures of grandchildren), titillation (e.g., pictures of hunky guys or swimsuit models), inspiration (e.g., posters of sports stars and fashion models), or give us a pure aesthetic experience (e.g., prints of the *Mona Lisa*).

*   *   *

Beauty can be partially explained by relatively simple evolved perceptual principles. But to understand the origin and our attraction to some of the stranger beliefs in this world, we need to look at how the mind can go wrong. For example, people can experience strange visions in the middle of the night. These visions can often be explained by sleep paralysis, an example of parasomnia, which is an often horrifying experience during sleep in which the victim is paralyzed, but fully cognizant of his or her surroundings. Vivid auditory and visual hallucinations, a feeling of pressure on the chest or of choking, and the conviction that an intruder is present often accompany sleep paralysis. Sleep paralysis happens to people in every culture. What is culturally specific is the interpretation of it. I'll mention only a few. In China, it's called "ghost pressure," in Japan it's thought to be a devil called a *kanashibari* stepping on the person's chest, and in Newfoundland, South Carolina, and Georgia, they call it the old hag. In the West Indies, it is Kokma, a baby ghost who jumps on sleepers' chests and attempts to strangle them. Similar ideas are found in Hungary and Indonesia. Beliefs surrounding sleep paralysis were so strong for the Hmong people that many of them died while experiencing it. In European history, the legend of the incubus and succubus demons involves one of them (usually) lying on top of the sleeper to copulate with them.

When undergoing sleep paralysis, the half-awake mind interprets the feeling of terror according to whatever cultural ideas fit best. This idea is similar to the attributionist theory in religion studies, which holds that many religious beliefs are interpretations of natural psychological experiences. If you've heard of the *kanashibari,* you might experience one when sleep paralysis gets you.

Contemporary Western victims of sleep paralysis tend to interpret it as an alien abduction. Often the tales of an alien abduction experience involve paralysis while in bed as the aliens enter the room. Most people who claim to have seen aliens admitted to being in the state between sleep and wakefulness when they did. The look of the classic abducting alien, the grey, was invented for a 1975 NBC movie *The UFO Incident.*[55] This image somehow became popular, with the result that now people in contemporary Western culture often interpret sleep paralysis as involving greys, often within the typical abduction narrative. This is an example of the bias called the availability cascade—in which ideas gain plausibility simply by being replicated in the media again and again—at work. People see pictures of greys, watch movies about them, and hear that other people claim to have been abducted. When they experience sleep paralysis themselves, their minds flesh it out with the story they've heard over and over.

This might explain why they see greys when experiencing sleep paralysis, but how did the greys become so prominent in the popular consciousness to begin with? I don't think it has to do with chance, but rather our perception of intelligence. In many ways, the greys represent a caricature of high intelligence, in the same way that Jessica Rabbit is a caricature of sexual female beauty.

These aliens supposedly have big eyes, large heads, small mouths, no noses, delicate fingers, and hairless bodies. These characteristics might seem arbitrary, but consider hypothetical aliens with the opposite characteristics: small eyes, small heads, huge mouths, big noses, fat fingers, and hairy bodies. Who would believe that such a stupid looking creature could possibly be super-intelligent (as the technologically advanced aliens presumably would be). The idea is laughable.

My student Meaghan McManus ran a study in which we presented people with computer-generated pictures of aliens and asked them to rate how intelligent the aliens in the pictures looked. Results showed that taller aliens were perceived as more intelligent than shorter aliens. Aliens with larger noses were perceived as less intelligent than aliens with smaller noses. Contrary to our prediction, eye size did not affect how the alien was perceived. Aliens who were tall and who had large eyes and small noses were rated as more intelligent than aliens who were short and who had small eyes and large noses.[56]

The point is that our concept of what these aliens look like, for the most part, corresponds to our ideas of what looks intelligent and what does not. It's the peak shift effect again, which happens when there is an exaggerated response to an exaggerated stimulus. Why are those particular features considered marks of intelligence?

My answer is that the greys look a bit like very young humans. For this answer to make sense, we need to explore the concept of neoteny.

In primates, the intelligence of a species is inversely proportional to how much their appearance changes from childhood to adulthood. For example, a study by biologist Mehmet Somel found that humans, the most intelligent primates, look much more like human babies than other primate adults look like their young counterparts.[57] In fact, baby primates of many kinds look rather more like human babies than those adult primates look like human adults.

When an adult looks a lot like the juveniles, scientists call the resemblance neoteny or pedomorphy. Learning ability is greater for the young. Perhaps the gradual increase in neoteny over the course of human evolution has increased our intelligence. Humans who develop faster than normal tend to have *decreased* cognitive abilities, and, as I mentioned in chapter 4, individuals who are less precocious at first end up being more intelligent later on. This suggests that making a primate species more neotenous might make them more intelligent.

It could be that the cause of neoteny, in mammals anyway, is the result of slowed development, as in the case of domesticated animals. Slowed development brings about a host of attributes to an organism. A beautiful example of this is the long-running silver fox breeding experiment being carried out in Siberia. By allowing only the foxes that willingly approach people to breed and disallowing the foxes that avoid people, the experimenters selected only for tameness. The surprising result is that, over the years, the tame foxes had other traits too that were not selected for, such as floppy ears, shorter legs, curly tails, shorter snouts, smaller skulls, and more juvenile faces.[58] They actually appear more like dogs. Humans might have bred dogs that resemble wolf puppies.

One theory of how this happened to humans, suggested by biologist Mehmet Somel, is that we domesticated ourselves.[59] As society grew more civilized, we wanted tamer people who were less prone to violence. We selected for them in a few ways. The most aggressive die because they get into deadly fights, but also because they are killed by other members of the culture as punishment. This effect tends to kill off 10 percent of the adult male population in contemporary hunter-gatherer societies.[60] You can imagine how rapidly a group could become nonviolent if the most violent 10 percent get killed every generation. The less aggressive people also might have been selected for sexually—mating with a killing machine makes less sense in a civilized society and more sense in a chaotic one. The result of these selective pressures might have been, just as in the case of the Siberian foxes, a slowing of development.[61] This theory suggests that slowed development accounts for our less violent behavior (relative to our ancestors), the way we look (more like human babies), and our increased intelligence.[62]

So how does neoteny explain why the greys look so intelligent? Greys look even more baby-like than humans! Just like babies, they are small, hairless (bald men appear more intelligent than men with hair[63]), have big eyes, and have a large head (relative to body size). I suggest that the apparent human neoteny of aliens is one example of what makes the whole idea of intelligent aliens compelling. Earlier I mentioned that greys might look like mothers as infants see them, and that this primal imagery might contribute to our finding them interesting. The fact that they also look like babies might contribute to their plausibility as intelligent beings.

* * *

A lot of this book is about how people believe nonscientific things for bad reasons. However, in the Western world, people tend to have a favorable view of science. In particular, they see mechanical explanations as having a certain authority that abstract explanations lack.

Cognitive neuroscientist Deena Skolnick Weisberg has shown, for example, that simply displaying neuroscience information when giving psychological explanations makes people more likely to accept those explanations, even if the information is not relevant to the explanation. The brain image, or whatever, gives the impression of hard, well-executed science.[64] Unfortunately, such tactics can be used in the courtroom. People tend to have a mystical idea of how the mind works and, consciously or not, see the brain as a somewhat unrelated entity. Showing a jury a picture of a brain might make them hold the suspect less responsible for an action. It's the "my brain made me do it" defense. If people are given a brain-based explanation for why someone did something, they assume that the person had no choice in the matter. In fact, all human choice is the result of some brain function. Whether or not people were in control of themselves has to do with more considerations than simply being able to find a responsible brain area or not.

*   *   *

We also seem to have a bias toward thinking that big, important events are caused by big, important things. This tendency was found in an experiment by psychologist Roy Spina.[65] It encourages the unwarranted belief in conspiracy theories when the generally accepted cause is judged to be too small and simple to be believable. For example, AIDS has had a devastating effect on many countries. The medically accepted story is that HIV, the virus that causes AIDS, jumped from nonhuman primates to humans. Its spread throughout humanity and the terrible cost of the disease leaves people in want of some kind of meaning to it all, some reason. Why would such a thing happen? Some look to divine intervention for meaning, and say that it is God's way of punishing gays and drug users.

Similarly, the attacks on the World Trade Center in 2001 appear to have been carried out by a small group of people. On the face of it, Lee Harvey Oswald acted alone when killing John F. Kennedy. These events, which had a great impact on the psyche of the American people, leave people in want of bigger, more complicated explanations. Conspiracy theories arise to fill this need. I am not claiming here that any *particular* conspiracy theories are false, only that, like many of the psychological effects described in this book, the fact that we have biases to believe in them means we should use extra caution when evaluating them.

*   *   *

People in the Western world have a psychological bias in favor of what they consider to be natural things, as opposed to manufactured or artificial things, particularly in terms of food or medicine. One study found that 58 percent of people preferred a remedy described as "natural," and when told that the medication was chemically identical to the nonnatural one, only 18 percent of them adjusted their answer.[66] When a food or medicine causes some bad effect, people have weaker feelings of regret and anger when it is natural than when it is synthetic. People also believe natural remedies are more effective, even with the same objective positive outcome. One study found that whether a medicine actually worked was of secondary importance to whether or not it was natural.

Why might this be? One reason might be that we tend to see in the news reports about bad side effects of artificial creations and that these are somehow more memorable. However, the news probably reports these items because they are more compelling, so this explanation has an element of circularity to it. Of course, a small bias in favor of naturalness can spiral out of control with the media. Although the reasoning is circular, so is the effect in the real world. It's a confirmation-bias feedback loop.

The real question is why we prefer natural things in the first place, in spite of the myriad natural dangers we all know about. After all, snake venom, poison berries, and hemlock are perfectly natural too. One possible factor is that artificial things are *newer*.

We seem to have a bias that makes us think that the longer something is in existence, the more valuable it is. This has been found for artistic works as well as for belief systems. In one study by psychologist Scott Eidelman people had more favorable views of acupuncture depending on how long they were told the belief system has been in place.[67] Perhaps this is based on the idea that the idea would not have lasted very long if it wasn't any good. This has bad ramifications for science-based medicine, which, as medical belief systems go, is relatively new.

It has been argued that we like "authentic" items more than copies because we perceive them to possess some special essence. Although looking at a photo or forgery of the *Mona Lisa* gives you much of the same *perceptual* experience as seeing the original, there is something about the *knowledge* that you're looking at the original that makes a difference. It somehow makes the experience more authentic, and authenticity is important to people. Perhaps it is in part due to the contamination biases we have. Leonardo da Vinci touched the actual *Mona Lisa*, not the poster sold at the college bookstore. I have a friend who can't go to art museums because she cannot resist reaching out and touching the paintings. One of the frustrating things about rocks from our moon is that scientists often have a hard time distinguishing them from the rocks in the parking lot outside of their laboratories. Would you rather have a rock from the moon or a rock from earth, even if there was no detectable chemical difference between them? If you are like most people, you would rather have the moon rock. There's a real importance placed on the authenticity of an object. Psychologists Bruce Hood and Paul Bloom found that even young children have this preference, preferring an object that was supposedly owned by a queen than a perfectly similar object owned by a commoner.[68]

Authenticity in a work of art is also important because people value a work of art more if they perceive that the artist has put effort into it. This would help explain why people prefer original works of art to reproductions and forgeries. It takes less effort to copy someone else.

*   *   *

Another interesting psychological bias people have is the "just world" phenomenon, which is the tendency people have to blame victims, based on the assumption that the world is a just place. The (largely subconscious) reasoning goes as follows: the world is a just place; some inexplicable tragedy befell an apparently innocent person; therefore, there must be something that person did to deserve it. The just world effect can make one look for reasons—a rape victim was "asking for it," for example. A study by psychologist Laurent Begue showed that people who did not give to a beggar were more likely to believe in a just world than those who did not, presumably because they thought the panhandler got what he or she deserved.[69] Fiction often portrays a just world, and studies show that the more people watch television, the stronger their just world bias is.[70]

One of the attractive things about many religions is that they promote the just world belief. It's sometimes all too apparent that, in the world we see day today, terrible things happen to good people and good things happen to bad people. This is disturbing. A belief system that promotes a view of the world in which everyone gets what they deserve can be pretty attractive. Religions step in to make us feel better about injustices. Indeed, people who are religious are less interested in any kind of wealth redistribution, perhaps because they believe the poor deserve to be poor.[71] Christian mythology tells of an afterlife in which the good are rewarded and the wicked are punished. In Hinduism and some sects of Buddhism, the good and bad you do in one life leads to a better or worse reincarnation in the next.

The belief in an afterlife is extraordinarily common, in spite of the fact that there is no scientific evidence for it. Psychologists have a fancy term for belief in an afterlife, *psychological continuity*.

Why is the belief in the afterlife so common? One reason is that it's kind of impossible to actually imagine what it's like to be dead. As far as science can tell, it's similar to a dreamless sleep, there's no consciousness and, well, no experience at all. It's not like *anything* to be dead. As cognitive scientist Jesse Bering points out, when we are dead, we won't even know it.[72]

Bering ran a study in which he asked people about the psychological states of dead people.[73] Most gave answers indicating that they were engaging in psychological continuity reasoning, and most reported that they believed in an afterlife. But surprisingly 23 percent of those who didn't believe in an afterlife answered questions in a way that revealed that they thought the dead had emotions, and another 35 percent had psychological continuity for knowledge, such as believing, remembering, and knowing.

Children have lots of psychological continuity reasoning and it tends to lessen with age, suggesting that it's innate, not cultural (if these beliefs were caused by culture and religion, you'd think that more exposure to culture, over time, would make the beliefs stronger, not weaker). This is on average; kids in Catholic schools were found to hold onto it longer. Luckily, it's not all innate. Wording things scientifically makes people less likely to think with psychological continuity, at least as indicated by what they say.[74]

Person permanence is a useful assumption in everyday life, but evolution might not have equipped us with being able to shut it off for people who are recently dead. Although people have no trouble understanding that bodies die, there is not much evolutionary pressure to get people to believe that minds die. So people go on thinking that minds continue to exist after death.

But if many people believe that humans survive their own deaths, then why is the death of a loved one so painful? One reason might be that even though we might believe in life after death, the communication channel with the dead is problematic at best. We might think we see signs from the deceased encoded in natural events, but it's not the same as sharing laughs and talking about your day in a normal conversation. So even if we think they continue to exist in some way, we still miss them.

*   *   *

We like to think that our tastes in the arts and the religious views we endorse are products of our character. But like our biology, the psychological biases we evolved to help us navigate our world and the cultures we grow up in are a hodgepodge of shortcuts and cultural idiosyncrasies that help determine that very character. The world puts limits on what these shortcuts can be, which constrains the human world of art, ideas, and religion. We like to focus on the differences between different cultures—their arts, their religions—but our underlying psychology requires them to occupy a rather restricted space of possibilities.

7

WHY WE GET RIVETED

I've described the basics of compellingness foundations theory, explaining, in turn, how each foundation of compellingness contributes to our understanding of the myriad things we find riveting. It's easy to cherry pick evidence to support these ideas, so I took care to report any counterevidence I discovered. Now that we have reviewed the foundations, we can ask again: why do we get riveted?

Culture affects preference. But I hope that this book has persuaded you that evolution has had *some* influence. Evolutionary explanations themselves are really compelling—people tend to give evolutionary explanations more credibility than they should: people find genetic explanations immutable, determined, and natural.[1] This is one reason many psychologists detest psychology based on evolutionary thinking—it's easy to come up with evolutionary explanations of behavior, and the media and public often will uncritically accept explanations that weren't scientifically tested. Because of this tendency we should exhibit extra care when evaluating evolutionary claims. Even if there is empirical evidence for a genetic influence, for example, we should ask *how much* influence is there? Answering that question is always complicated. Because many of these explanations involve evolutionary reasoning, it's important to understand the complex ways evolution has affected humans and human behavior. The idea that evolution simply affects our genes, which in turn affect behavior, is too simple. The truth is more nuanced—culture can affect genes, as when the cultural tendency to drink milk in northern Europe resulted in people genetically changing so that they could digest milk into adulthood. It was the culture's desire to drink milk that made the ability to digest it a force of natural selection.

I have discussed the arts mostly as a by-product of other evolutionary adaptations. That is, we like art because it satisfies desires that were evolved for something else. But it does not explain particularly well why people are driven to *make* art.

The creation of art involves expense. It requires free time, material resources, energy, hand-eye coordination, intelligence, creativity, and occasionally risk. It could be that art serves as an indicator of an adaptive trait, much like the male peacock's tail is an indication of its health and ability to acquire resources, or the reason we find lustrous hair on a woman attractive. They are called "fitness indicators." The hair itself might not help you survive, but it might help you reproduce. Similarly, it could be that the artistic ability of a person informs mate choice, as believed by every bassist who joins a band to pick up girls. Artistic ability predicts access to resources, and to some extent it also predicts intelligence. The higher your IQ is, the more creative you are, but only up to 120: having a higher IQ than this does not predict *even more* creativity.[2]

It is also an intriguing thought that artistic production might be an evolved trait in itself, and that art might be an "extended phenotype," much like a beaver dam or a bower bird's bower. Another possibility is that we have an instinct to create art to increase group cohesion, an argument put forward by anthropologist Ellen Dissanayake. Art that is communally made can draw people together. Interestingly, psychologist Jonathan Haidt and others have argued that religion evolved for the same reason.[3]

* * *

What can the foundations of compellingness tell us about celebrity worship? Our fascination with celebrities—be they politicians, athletes, or artists of various kinds—is a curious phenomenon. It is clear that in arts and sports, we have a love for virtuosity. One reason is that when we see virtuosity, we see something rare, perhaps something we've never seen before. We remember unusual things better.

Second, we might find celebrities inspiring. One study by psychologist Jaye Derrick found that psychology students with low self-esteem felt closest to those celebrities who were similar to their ideal images of themselves.[4] In our evolutionary history, highly variable environments favor an imitative strategy, because genetic evolution and learning how to do things all by yourself requires too much time. Imitating successful people is a great strategy, even though it can lead to perplexing fads and weird haircuts. Perhaps this is the reason there appears to be more celebrity worship in the young. They have more to learn and thus are drawn more to role models. This also explains why we like to see incredible failure. Rubberneckers have always looked at accidents, and in the Internet age "fail" videos (often of people getting hurt doing something dangerous) are very popular.

A third, related hypothesis that would apply to displays of physical skill is that we find it pleasurable to imagine ourselves doing what we are watching. Recall that, when watching dance, the motor areas of the brain are active. It could be that when we see someone do something amazing, we are, in our minds, doing it ourselves, and this is thrilling.

Fourth, our love for celebrities might be another ramification of positive contagion. Maybe we think that something good will "rub off" on us by being close to greatness.[5] Of course, we observe most virtuosity on screen, but because most of our brain cannot tell the difference between representations and reality, the same greatness detector goes off in either case.

A related concept is that of "costly signaling." The idea behind this is that obvious impediments to your survival are attractive because you must be fit indeed to be able to survive with such a handicap ("If he can live with that impediment, he *must* be strong!"). The peacock's tail is a great example of this—if a male can grow a huge, healthy tail in spite of the fact that it makes him more vulnerable to predators and requires costly nutrition resources, he must be a good peacock to father the peahen's chicks. This is the handicap principle. Likewise, it could be that personal expense is related to our admiration. Just as conspicuous consumption signals wealth, showing that you have the time and resources to make a costly piece of art can show fitness for reproduction.

Fifth, we might perceive the people we admire as our leaders or we might, subconsciously, want to make them our leaders. If we perceive that we are allied with those we watch, we might feel that our "team" is winning. Male sports fans get a surge of testosterone when their favorite sports team wins, according to a psychological study by Paul Bernhardt.[6] A related concept is kin selection. We might feel particularly allied with people with whom we have a genetic overlap. This hypothesis would predict that people enjoy watching virtuosity in people who appear to be like us (e.g., the same race, hair color, or build) more than virtuosity in people unlike us. This is the same reason we are proud of our children when they do something wonderful—it feels good to us because it is an indicator of survival and reproductive fitness. To a lesser degree, we should have more care for nonfamily members who share more of our genes than for nonfamily members with whom we have no genetic overlap.

An interesting question is why we often do not feel the need to compete with people who display virtuosity. Recall the relative social status hypothesis: we will not feel in competition with people who are much higher or lower than we are in the perceived social hierarchy.

The relative-social-status hypothesis also predicts that, as in gossip, news stories about other people will be most compelling when they are about people lower than we are (on the social ladder) improving their lives, or about people higher than we are having some downfall. Poor people getting poorer appeals to our sense of danger, but not to our interest in our own place in the social hierarchy.

Many of us are hooked on the news. News agencies know that putting a "human face" on a "story" makes it sell. This practice reflects our desires to know about other people, preferably in some narrative form. Even stock market changes will be cast as "agentive," that is, as though the stock was an agent capable of making decisions.

Should people follow the news? Some activities are pleasurable and doing them is inherently rewarding. Others are difficult to do, but pay off with happiness or pleasure in the future. Some other things are important to do, even if you never get much from it. I worry that news has none of the characteristics that make something worthwhile. It's not fun, it causes anxiety, it gives you a warped sense of reality, and people who watch it are rarely going to *do* anything with the information they get. For most people, watching the news is like sharpening a saw that they will never use to cut anything.

So why do people watch it? Its sensational nature makes it feel important when it's really not.

News reporters are often good storytellers. If someone is an excellent writer or a superb speaker, our guard will be down. Beautiful ideas are not always true, and when we encounter a compelling idea, we must take care.

The flip side of this is that just because someone is a terrible writer or speaker does not mean that he or she is wrong. This is a particular problem with nonnative speakers of any language, whose often poorer communication skills make their ideas seem unworthy. Use extra caution with the talented and be more generous with those lacking in rhetorical and linguistic skills.

Just as familiar ancient myths are known today because they were good enough to be copied and retold again and again, religious stories, ideas, and practices also survive because they are compelling. One of my goals of this book is to explain why we find religious and paranormal ideas riveting. It is my belief that supernatural beliefs are false, and the science backs that up. But some skeptics would have you believe that no well-conducted scientific studies have ever found evidence of the paranormal. This is wrong for a couple of reasons. First, anything that is repeatedly found to exist because of scientific studies is no longer considered paranormal, for example, hypnotism. Second, even for some paranormal phenomenon that

does not exist, some percentage of studies will find statistically significant results nevertheless. This is an unavoidable consequence of
using statistics in science.

There is variety in all psychological measures. This is why we
use statistics. The scientist might have gotten a sample that was not
representative of the group. For example, *some* children know more
than *some* adults. How does the scientist know that her study, which
found that adults know more than children, wasn't due to having a
weird sample, or, in other words, due to chance? When you hear a
scientist report that there is a "significant" difference between two
groups, it means that the differences observed are unlikely to be due
to chance. Scientists set a threshold, say 1 percent. In this case if the
statistics show that there is a 99 percent probability that the effect
observed was *not* due to chance, then the study has "significant"
results.

However, this means that one out of every hundred studies
conducted will have significant results even if there is no effect at
all, because the scientist set a 1 percent threshold for significance!
This happens for the same reason as why a test that is 99 percent
accurate will give an inaccurate result 1 percent of the time. To
return to the paranormal, for every 100 well-conducted studies on
clairvoyance (the ability to "see" what's not visible with the eyes),
we would expect one of those studies to come out with results that
support clairvoyance. If a scientist runs 100 studies, she will likely
only submit the one that found significance for publication, and
never submit the rest—this is science's "file drawer problem."

The file drawer problem is amplified by the publication bias of
journals to publish only significant results. So what ends up happening is that the few paranormal studies that actually find significance
(the 1 percent that were due to chance) get published and the ones
that fail don't get accepted by the journal—if they get submitted at
all. So if you read the literature, it might look like the paranormal
is real!

All that is to say that the existence of a single study is not suf-
ficient to prove that some effect is real (this applies to all science, not

just those studies investigating the paranormal). What is needed are studies that find the same results over and over, and can be replicated by other scientists. Studies of paranormal phenomena fail in this regard, which is why scientists don't believe in the paranormal.

Just as paranormal ideas have foundations of compellingness working in their favor, we have a great many biases working against our believing the scientific alternatives. Many well-supported scientific theories violate common sense. They might violate the common beliefs of a particular culture (one society believes that kids won't learn to walk unless they are buried up to their waists in a standing position[7]), but also might violate things *everybody* believes. This is particularly true for things that are not medium sized (in this context, anything bigger than a flea but smaller than a planet is considered medium sized. Nonmedium-sized objects work according to rules that dominate at subatomic, atomic, or astronomical scales). Elementary particles can be in two places at once. Events happen without cause. Such things are hard to swallow, even for physicists. I still have to remind myself, watching a sunset, that it is the earth that's moving, because we can't feel the motion.

We evolved to make sense of the world we were in, and what I mean by "world" is what Richard Dawkins calls the Middle World, the set of objects we directly interact with. Our solar system is not a part of the evolutionary world of our ancestors in the same way that a forest is not a part of the smaller world of an earthworm. In both cases, these larger worlds were not relevant to the business of survival and reproduction. We evolved to understand the regularities in our world, and only recently (and by recently here I mean the last few thousand years or so) have we tried to understand things outside of the world we live in day to day. Unfortunately for democratic societies and science education, this world beyond—the very big, the very tiny, the very slow or fast—can be a little weird. Our minds just are not set up to understand it.[8] Our intuitions about many things end up being wildly off. For example, imagine that our solar system was shrunk down so that the sun had the diameter of about half a meter (about the length from your elbow to your fingertips on one arm). At this scale, how far away from the sun would the earth be, and how big would it be? Take a guess before looking to the note for the answer.[9] If you're like most people, you find it very hard to accurately estimate very large and very small distances. Things we don't understand well don't "stick."

Not "sticking" is just one of the problems that all ideas, religious ones included, must face in the marketplace of ideas. Another is simply communication—successful ideas have to be able to be efficiently communicated from one person to another over time. This is only possible with the successful religions of the world because they take advantage of systems of understanding that our minds already have. In religions the world over, gods and other supernatural creatures are often depicted with fangs and teeth, and have the power to protect or destroy us. Thinking of a god and people as being in a predator-prey relationship is something that makes sense to us; we don't need to read whole books to grasp the basic idea.[10] We hear brief descriptions and basically get it. Not so with quantum physics. Our minds, based on brains that evolved to understand what we can see and touch, fight to reject scientific explanations that violate our common sense, opening the door to false explanations that better fit with our preconceptions.

*   *   *

Although belief in the paranormal and in traditional religion is not the same thing, a survey by psychologist Karl Peltzer showed that belief in one often predicts belief in the other.[11] Religion is an extraordinary thing. Every culture has it. Even if you don't like the word *religion* and think it's too vague to mean anything, as anthropologist and cognitive scientist Scott Atran says,

> In every society known, there is: 1. widespread counterfactual belief in supernatural agents (gods, ghosts, goblins, etc.) 2. hard-to-fake public expressions of costly material commitments to supernatural agents, that is, sacrifice (offerings of goods, time, other lives, one's own life, etc.) 3. a central focus of supernatural agents on dealing with people's existential anxieties (death, disease, catastrophe, pain,

loneliness, injustice, want, loss, etc.) 4. ritualized and often rhythmic coordination of 1, 2, and 3, that is, communion (congregation, intimate fellowship, etc.).[12]

The academic discipline of religious studies is dedicated to understanding religion. Its value, and indeed whether or not the very existence of religion is a good or bad thing for the world, is a current matter of great debate in religious studies.

Trying to explain why we have religion is difficult in part because many people believe that they already know the answer (typically, people think it's because it makes people happy or provides explanations for life's mysteries). They suffer from "premature curiosity satisfaction."[13] Both of these oft-mentioned explanations have truth to them, in many cases, but it's much more complicated than that.

We are attracted to religion for many of the same reasons we are attracted to the arts and other ideas. But aren't there differences between fictional characters and gods, for example? Atran wants to know why some people believe in Vishnu but nobody believes in Mickey Mouse.[14] This is an important question, but I think it's telling that at one time many people believed wholeheartedly in the god Thor, but now he's basically Mickey Mouse—a cartoon superhero. What made Thor compelling as a god in the past also makes him a compelling cartoon character in the present.

There are differences, of course, in that there are lots of people who actually believe that Vishnu exists, but (I would hope) nobody believes that Mickey Mouse exists. Atran believes that cognitive theories cannot "in principle" distinguish the two, but it seems clear to me that some of the benefits of religion (providing explanation, comfort, etc.) only come if people believe in it. Muslims might find it interesting that some Mormons believe that they will become gods of their own planet after they die.[15] For nonbelievers such a belief *is* interesting in the way that Mickey Mouse is. But you will not find this a *hopeful* thought for your future unless you actually *believe* those ideas. A cognitive reason for why Mickey Mouse is different from the belief in angels is that believing in Mickey Mouse does not appear to confer the same benefits to a person that believing in an afterlife does.

When you indicate that you believe a statement, religious or otherwise, the ventromedial prefrontal cortex is particularly active, according to neuroscientist Sam Harris. This area is associated with emotions, rewards, and self-representation. This is the same for Christians and nonbelievers, supporting the idea that religious beliefs are believed the same way as other beliefs are.[16] Indeed, the causes of peoples' beliefs in religion are also the cause of other beliefs, be they scientific or superstitious.

The majority of people in the Western world believe in God. Even 40 percent of scientists are believers.[17] One of the most striking things about religious belief is its pervasive anthropomorphism. That is, religion tends to personify things that modern science sees as impersonal objects or forces. This is explained by the social compellingness hypothesis, which holds that we have a built-in desire to find social relationships important. Explanations couched in terms of conflicting personalities are compelling for the same aesthetic reasons that good stories are compelling. They resonate with our minds, and we are more likely to believe them.

Even beliefs that do not personify natural forces often have something directly to do with people, either in their ability to heal us, hurt us, reincarnate, etc. We tend not to have religious beliefs about things that have nothing whatever to do with human beings. In contrast, there are lots of scientific theories that have nothing to do with people.

The autism spectrum quotient (AQ), which I discussed in chapter 1, is probably a decent indicator of how much you think in terms of people and their relationships. If you have a low AQ, you are probably more likely to believe in gods and other supernatural beings, and use them to make sense of the world around you.

In a way I feel bad for humanity. We are wired to relate everything to ourselves, and science has shown that the vast majority of the goings-on of the universe have nothing whatever to do with human goals. As science understands more of our universe, it whittles away at human prominence in the grand scheme of things.

Our desire to believe what we hope is true makes our minds fertile soil for ideas that allow us to believe that death is not the end. It helps us deal with our sorrow at the loss of loved ones, and also any anxiety we might have about our own deaths. But not all visions of an afterlife are rosy. The belief in ghosts is usually rather negative, as is the ancient Greek vision of afterlife in the underworld. However, many afterlife beliefs are relatively positive and involve a sense of justice (such as Christian ideas of heaven and hell and Hindu ideas of reincarnation) that further draws us in, because justice makes us happy. We like to see the good rewarded and the bad punished, and too often affairs in the real world don't work out that way. We should not let our hope that the world is just interfere with our compassion.

We appear to have an innate idea of psychological continuity, the belief that certain mental states persist after death. This might be a side effect of person permanence, which allows us to know that people do not cease to exist when they are not physically present and are "out of sight." These characteristics might have been a factor in the origin of beliefs about the afterlife.

We know some things about the areas of the brain that are involved with various forms of religious experience. The left hemisphere, for example, appears to tie our experiences into narratives, representing experience in terms of languagelike representations. When the left hemisphere is quieted, by, for example, a stroke or meditation, we can consciously experience the right hemisphere's view of the world, which is more holistic, contextual, and uninterpreted. The quieting of the parietal lobe appears to have similar effects—feelings of being one with the universe and a loss of a sense of self.

Many of the so-called miracles of religious history would be considered psychotic experiences today. The Buddha (or his followers, anyway) claimed he was tortured by the demons of Mara the night before his enlightenment. Jesus emerged from the desert after, supposedly, meeting with Satan.

There are drugs (and perhaps Koren helmets—see chapter 5) that can induce experiences that are often interpreted as having religious significance.[18] As tempting as it might be for a skeptic to view these findings as evidence of the illegitimacy of the metaphysical beliefs supporting these interpretations, the view is ill advised. Does finding the brain areas associated with revelatory experiences mean that interpretations of them as supernatural are incorrect? No. Interpreting the findings that way would be to assume that any experience of a god must not involve our brain, which is a hard position to defend. There are brain explanations for why we see peanut butter, but from that we would be unwise to conclude that peanut butter doesn't exist. Just because I can make you experience light by pressing on your eyelids in a dark room does not cast doubt on the existence of light when perceived normally.

However, we should doubt certain interpretations of experiences if we have reason to believe that those interpretations are caused by something evolved or learned for something else entirely. When you talk on the phone to an automated system, you might get frustrated and angry and raise your voice when speaking to it. Talking on the phone is a social communication intended for dealing with other human beings, but we are applying it to a very simple computer system that (at the time of this writing, anyhow) cannot usefully perceive a caller's anger. We act like we are speaking to a real person because we're talking on the phone. What is happening here is that our perceptual system, evolved to recognize what it sounds like to talk to someone, is being activated by a machine. In this case, the perception is wrong. There is no person there. Likewise, if the reason we are interpreting a revelatory experience as supernatural is being caused by, say, an overactive theory of mind, we would have reason to doubt an interpretation of that experience as involving a god.

What these ideas also cast doubt upon are interpretations of some individual experiences. If someone has a revelatory experience that was caused by drugs, we can doubt interpreting it as being religiously significant for the same reason we doubt our natural interpretations of optical illusions and hallucinations. As the atheist joke goes, believing your own hallucinations is mental illness; believing other people's is religion.

Personally, I believe that even these experiences for which we know of no physical cause do not warrant religious interpretation. We should not take religious experience as indicating something fundamental about the nature of the universe as a whole. We should take religious experience as telling us something about *us*.

People can have strange experiences, possibly induced by meditation, singing, ritual, or drugs, and people interpret those experiences according to their beliefs and culture. They misinterpret a psychological phenomenon as a divine experience. For example, the near-death experience appears to be a cross-cultural phenomenon, although the content tends to differ according to culture. Religious individuals are more likely to have spiritual interpretations of it. This supports the attributionist theory of religion—that when we undergo a strange experience, we interpreted it in a way that fits with our cultural beliefs. Of course the person undergoing the experience is unaware of this and uses the experience as further evidence for the beliefs he or she already has. One group, the Emil Society, even interprets falling asleep during church lectures to be a sign of religious experience![19]

What's happening can be understood with the old and new brain distinction. Unconscious processes "believe" one thing and at the same time your more logical processes might "believe" another. When you find yourself in this state, there's no reason to worry that you're being contradictory. When you look at an optical illusion that appears to be moving even when it's not, you're in the same state— your old visual processes detect motion; your new processes know it's not there.

Experiences of possession and trance are partly due to brain malfunctions that disable our conscious connection to the motor control of our bodies. This might happen in schizophrenia, dissociative identity disorder (multiple personalities), and in somatoparaphrenia (denying the ownership of a limb or entire side of one's own body). Often interpretations of these experiences are informed by one's culture: hearing voices means an FBI chip is implanted in the head, or partial paralysis is attributed to a demon possession (this interpretation is called the delusion of passivity). There is also evidence suggesting that people report possession to describe shameful and all-too-real experiences such as rape.

Just as many metaphysical religious beliefs might have originated with the ideas of schizotypals, many of the rituals we see in religions might have started with people with obsessive-compulsive (OCD) spectrum disorders. Orthodox religions are replete with food and body cleansing, repetition of mantras, numerology, and rules about how to enter and leave places. It could be that people with OCD-spectrum problems become religious leaders: leaders are often the most fervent followers of religious rituals. Martin Luther might be retroactively diagnosed with some variant of OCD, having written "the more you cleanse yourself, the dirtier you get."[20] However, it doesn't matter what religious belief or ritual somebody makes up if it does not conform to people's preexisting tendencies to find certain kinds of things riveting. Some ideas just don't take.

People who have tendencies toward obsessive-compulsive symptoms will be drawn to religious traditions and practices that dovetail with their symptoms, as evidenced by the fact that Catholics tend to have more OCD symptoms than Protestants.

It might sound implausible that people would listen carefully to and follow the religious ideas of the mentally ill. But their strange behaviors can be interpreted as divine intervention. Take, for example, work from anthropologist Harvey Whitehouse, which describes the case of Wapei, a man in Papua New Guinea. He experienced a violent trembling fit and had a vision of Jesus returning to his land, along with Wapei's ancestors (reincarnated as Caucasians) bringing Western goods with them. This experience resulted in a short-lived cult called Noise. People believed him not so much because of what he said but because of the trembling that accompanied it.[21]

Another Papua New Guinea man started gesturing as though he could not speak. The others interpreted his gestures, encouraged by his smiles, as religious revelation. This was the beginning of yet another cult, this one called the Ghost cult. We know that it was his apparent inability to speak that sparked it, because the things he was communicating were not significantly different from the things the community had heard every day in religious services.

People often don't understand mental illness or the behaviors that accompany it. They confabulate explanations, and it's easy to confabulate a religious explanation when mental illnesses so often go hand in hand with religious experiences.

The circle becomes vicious, because the very contents of people's hallucinations contain elements that are compelling to the rest of us: primal imagery such as the dead and snakes, and feelings of power and moral certitude. New religions don't get started based on Pokémon or cars.

The same is true with ritual: the rituals that the OCD-spectrum religious originators came up with match pretty well with the day-to-day OCD tendencies many people have. They resonate with our minds. They're compelling. People claim that their religious rituals give life meaning.

Temporal-lobe epilepsy is another brain disorder that might have been a factor in the origin of certain religious beliefs and practices. People who have it tend to be solitary, humorless, and rigid in their ways, and shy away from new experiences. They also tend to have hypergraphia, which is a compulsion to write, and an interest in religious subjects. Although such people might not be particularly religious, hyperreligiosity is a feature of some types of temporal-lobe epilepsy. Their seizures involve seeing colors and an intensified emotional experience, and sometimes euphoria, particularly with religious words and imagery. Stimulating the temporal lobe can cause a seizure or trigger hallucinations of paranormal phenomena. The experience of epilepsy can cause permanent changes to the temporal lobe. One can imagine how someone interested in religion and with a compulsive tendency to write could interpret the seizure as a religious experience and write about it, possibly changing the course of religious history. In fact, some neurologists suspect that Saint Paul himself had temporal-lobe epilepsy.[22] Hallucinations from schizophrenia and temporal-lobe epilepsy might have inspired aspects of Islam and Mormonism in historical figures such as Muhammad and

Joseph Smith. Drug-induced hallucinations are a part of many Native American traditions. There is reason to think that Joan of Arc also had temporal-lobe epilepsy.[23]

Some characters in myth and folklore might be based on archetypal personalities related to mental problems. For example, attention-deficit/hyperactivity disorder (ADHD) might be the origin of the trickster character that seems to turn up in stories around the world (e.g., Anansi, Br'er Rabbit, the coyote).[24]

I've described how people with mental illnesses might have been implicated in the origin of religions and how certain mental illnesses might be a factor in making or keeping someone religious. But we don't need to go that far. Altered states of mind, not extreme enough to be considered an illness, play a crucial role in many world religions.

Whitehouse compared religious beliefs and practices indigenous to Papua New Guinea to the Christian practices in the same place after missionary involvement.[25] He differentiates ways of spreading and maintaining religion as either "doctrinal" versus "imagistic." You are probably familiar with the doctrinal way, because it is the primary means used by Christianity, Judaism, and Islam. It is characterized by sacred texts, a logical explanatory structure, beliefs that can be expressed and communicated with language, and repetitive, often daily, ritual. Sound familiar?

Religions that focus on the imagistic way have very infrequent, emotionally powerful rites. These rites are often horrifying, and can involve beating, years of isolation, pain, and sleep deprivation. Some people do not even survive them. These rites of terror result in realizations and experiences about the nature of the world that are difficult, and sometimes even taboo, to talk about with other people. In precolonial Melanesian fertility cults, for example, talking about the meanings of religious symbolism was punishable by death. These experiences and the meanings they generate for a person are unique, and often are remembered for a lifetime. You just have to experience them for yourself to understand.

Contrast this way of learning about one's religion by sitting and listening to someone read from the Bible week after week, which relies less on a powerful experience and more on an intellectual understanding of the material. Although science is often pitted against religions like Christianity, science and these religions are similar in that they both feature knowledge structures that endeavor to be internally consistent, and both provide lots of answers delivered through language. This is quite different from gaining an understanding imagistically of, say, the relationship between the pigs, the birds, and your people that you got during an unforgettable rite of terror you experienced when you were twelve years old, one that involved mystery, extreme sensory stimulation, and mortal danger. It's not that religions like Christianity are purely doctrinal, it's that the doctrinal way is the most common method they use. Even primarily doctrinal religions can enter phases of imagism, as certain charismatic churches do when people speak in tongues.

*   *   *

Religions can be thought of as sets of ideas—about the nature of the universe, how things should be done, etc. Like other aspects of culture, religious ideas spread, or don't spread, based on a number of factors.

Because the religions we tend to hear about are the ones that are the most popular, it is easy to think that they are representative of religion in general. The religions people are most familiar with are Christianity, Judaism, Islam, Hinduism, and Buddhism—the "big five." However, there are far more than five religions in the world today. Depending on how finely you categorize them, there are hundreds or thousands (one estimate is 4200), and many, many more if you look over the course of human history and include religions that are no longer practiced.

There are too few comparativists in religious studies. The top five religions are so huge that they describe families of sects that we further subcategorize. Nevertheless, the fact that religions not in the top five get studied so much less is a great disservice to the scholarly pursuit of knowledge about religion.

The focus on the big five is a mistake, perhaps most crucially because characteristics of most religions *prevent* them from becoming wildly popular, resulting in a skewed view of religion in general. One reason is that all the top five are primarily doctrinal and in that sense similar. For a religion to spread all over the world, it must have special characteristics of its own, and these characteristics make it unlike the less popular religions and to that extent uncharacteristic of religion in general.

Because imagistic means are individualized, as a religion spreads it can change greatly from town to town. One of the reasons why doctrinal religions have spread so successfully is that sacred texts can be copied with high fidelity, and intellectualization and language-based practice can enforce uniform interpretations. Indeed, such religions often actively stamp out religious innovation, sometimes through exile or excommunication of the innovators. This powerful combination helped primarily doctrinal religions spread across the world in a way that primarily imagistic religions could not, since, without doctrinal reinforcement the religion would change so much as it passed through different cultures and conditions that it would become unrecognizable as the original religion. This is why all the major religions of the world have scripture, and why the layperson would be forgiven for not even knowing about the ones that did not.

Sociologist Max Weber noticed that even staid doctrinal religions have outbursts of charisma, with a return to primal feelings and evocative ritual involving trance, revelation, and so on.[26] Doctrinal religions try to keep heretical outbursts from spreading, because nondoctrinal information is hard to codify and control. But codified, controlled information can be boring, increasing the chances of imagistic phases. Such oscillations have been documented for Islamic, Hindu, Christian, and Buddhist movements.

Many religions have supernatural ideas about very specific things, such as species of plants or animals, or particular mountains. For example, there is a Sudanese cult that believes that a certain species of tree, the ebony tree, can overhear conversations conducted in its shadows, and that these conversations can be extracted with particular rituals.[27] Think about how this might go over in places without ebony trees. Likewise, if a particular mountain must be climbed, then followers of the religion must be within a certain geographical distance. This too would hinder the spread of the religion. (Islam is rather exceptional in this regard in that it requires a pilgrimage to Mecca. However, there are caveats. The pilgrimage is not required for people who cannot afford it and it is required only once in a lifetime.)

As a result, I would expect that popular religions would culturally evolve to exclude these geography-limiting factors. If they did not, they would be outcompeted by other religions. My theory predicts that the most popular religions would have fewer tenets involving species and location-specific things.

Compelling religious ideas spread for the same reasons other compelling ideas spread. The ideas resonate with people on a deep level. They get repeated and communicated through public discourse. In modern times, mass communication spreads ideas very rapidly. Repeated exposure to their ideas makes them seem more plausible, due to the availability heuristic. When we're surrounded by people with the same beliefs, we tend to agree with them, due to the bandwagon effect and the herd instinct.

Once ideas are believed, we seek evidence to support those beliefs and interpret ambiguous things we see as supporting evidence. For example, we might see God as responsible for helping someone recover from a disease, but not as the cause of that disease in the first place.

* * *

Recall that evolution (by natural selection) occurs in any medium that has variation, heritability, and selection. New religions constantly pop up. People and groups of people explore the space of possible religions in this way. Thus the variation criterion is fulfilled. Religion is spread by enculturation, usually from parent to child, that is, it is replicated. Most religions, however, fade away. Only some last. That is, they are selected.

It's clear that people tend to take up the religion of the society around them. However, societies vary regarding how *many* religions are available to their constituents. In countries where there are lots of religions (or branches of the same religion), people can choose the one that's best for them. With more choice, people should be more likely to be religious. After all, if they don't like the dominant religion, they can go next door to another one. And this is exactly what happens. In countries with more religious "competition," people tend to be more religious. Monopolies impede religious markets just as they do nonreligious ones.[28] It should be noted that it's not clear from these particular findings whether religiosity causes competition or vice versa. The freedom of religion that Americans enjoy is one explanation that has been offered for why Americans are more religious than their European counterparts—it's not that there's something fundamentally different about Americans, it's just that there is more religious choice in the United States.

If there is evolution happening for religions, we would expect that, just as in animals, certain traits that are good for a religion's survival would be common in successful religions. There are other traits, such as the prohibition of having children, which are so bad for the replication of a religion that no successful religions have them.

Various scholars have come up with some good hypotheses regarding what some of these traits of successful religions are:

1. The inclusion of a belief that one should not speak with anyone outside the religion. This is typical of cults. If you speak to people who are not in the cult, you are more likely to get talked out of it. A related prohibition is not to marry outside of the faith, which compromises the religion being passed on to children. Some Jews are explicit about this purpose—some justify the prohibition of marrying gentiles on the grounds that their religion's survival depends on it.

2. The reliance on revealed experience and on the testimony of those who have had revelations.[29] If the religion makes

claims about objectively observable reality, it can be more easily falsified. It's hard to argue against revealed experience.

3. The cultural norm that it's rude to challenge someone concerning his or her religion.[30] This uses social pressure to keep people from discussing the subject.

4. A prohibition against criticizing the religion.[31] One good trick for a religion to take up is to inoculate itself against critical thought. If the religion can keep people from thinking skeptically about the religion, it has a better chance of survival. Judaism appears to be a notable exception.

5. An emphasis on faith. Faith is, by definition, belief without rational justification, a nonrational justification for belief. Some religious people have gone so far as to discourage rational argumentation for God's existence because it undermines the principle of faith. If a people reject rationality, there can be little serious *discussion* about what's correct and incorrect. Dogma fills the gap left over.

6. Prohibition against defection from the religion and against simultaneous practice of another religion. It's clearly to a religion's advantage to discourage its members from leaving the faith, and some will threaten their members with harm, either in this world or the next, for doing so. A related notion is the belief that those outside the faith are bad people (e.g., "the wicked," or "infidels"). Believing in two religions at the same time is dangerous because it introduces a competition and comparison of the two faiths—in addition to outright contradictions.

   I would suggest that the wilder and less commonsensical are the teachings of the religion, the more likely that religion will be to have the above traits.

7. Unfalsifiable and incomprehensible claims.[32] You can't find evidence against the existence of a deity that leaves no measurable trace on the observable universe, nor against claims that are so open to interpretation that one cannot get unambiguous predictions from them.

8. An appreciation for mystery for its own sake.[33] If mystery
   is appreciated, adherents will not be so keen on trying to
   resolve the mystery. This serves the same purpose as prohib-
   iting critical thought.

*   *   *

There is a popular belief that religion is a top-down kind of organization. That is, that people get their religious beliefs from religious
leaders. But religious leaders (we'll call them priests for purposes of
this discussion) as we know them are, historically, relatively recent.
Studies by anthropologists Stephen Sanderson and Wesley Roberts
show that societies that do not have writing, that are small, and
that use hunting-gathering to get food tend to not have priests, exactly, but shamans, who are distinguished from priests in that they
work part-time, often for a fee, and are not considered the final
authority on the supernatural world.[34]

Societies tend to change in a predictable way, going through the
same steps in the same order: from egalitarianism, to the differentia-
tion of social status, to having full-time craft specialists, to economic
markets, to legal codes. Similarly, as societies change, their religions
tend to change too. The order tends to be: shamanistic, communal,
polytheistic, and finally monotheistic.

Shamanism is characterized by having shamans as the center of
religious practice and conduit to the supernatural, a belief in animism (that objects and animals have spirits, and perhaps that people
are descended from animal totems), and a lack of calendrical rites
(having to do with seasons). Sixty-two percent of shamanistic societies appear in foraging societies, and 90 percent of them in places
with no writing or record keeping. [35]

As societies begin to farm, they tend to enter the communal religious phase, characterized by calendrical rites and groups or people
(perhaps in addition to shamans) being conduits to the supernatural. This is most common in agricultural societies, and 93 percent
of them are in nonliterate societies. The idea is that farming brings

about anxieties that hunter-gatherers do not have and that religious rites are created to attenuate them.

When a society gets bigger, it enters the polytheistic phase, with priests and a pantheon of distinct gods. The priests are often in alliance with political leaders.

At some point in the polytheist or monotheist stage, societies become literate. Monotheism is like polytheism, but with a single, usually all-powerful, god. Monotheism is most often found in literate societies with intensive agriculture that uses metal plows. The impact of literacy is great—as the sacred texts get more complicated and incomprehensible over time, the priests are "needed" more and more for interpretation. This monopolization of doctrine helps the priestly class gain power and influence. Priests are only possible in societies rich enough to support people with a full-time religious occupation. The domination of the world by monotheistic religions is a fairly recent phenomenon—only since 600 BCE. There is no agreed-upon explanation for this "monotheistic great leap forward."

But even in societies with religious leaders, it might be that religious authorities, being more educated about science, will be less prone to some of the more extreme claims of religions. However, religious authorities cannot simply make their followers believe anything they choose. Because the universal themes of religion are innate, religious leaders must temper their intellectual claims to match what people are expecting or else they will be ousted. Perhaps, if people listened to their educated religious leaders more, some kinds of supernatural beliefs would diminish over time.

As we have seen, there are interesting regularities across the religions of the world. Some might see this as a kind of common denominator of an underlying reality, as evidence of the actual nature of the supernatural world. Another way to look at the patterns is that we are all human, with human minds, and the religions that spread and survived are those that our ancestors, and we today, find compelling.

Art and religion work on the mind in similar ways. This is why it is not surprising that religion often uses art explicitly (paintings, songs, dances, etc.) and that we often use old religions that nobody believes in anymore (such as ancient Greek religion) as art. The very stories that the ancients believed were real we now read for entertainment.

The fact that religion changes over time indicates that it, in some sense, "evolves." But this does not mean that we have a genetic predisposition to have religion, nor does it mean the even stronger claim that religion has been adaptive in our evolutionary history and has been selected for.

*   *   *

Science and religion have two important things in common.

First, they have generated beliefs that people endorse or reject, and the beliefs of science can be in conflict with those of religion. The facts you learn in school and hear on the news are what most people consider to be "science." We can say that science and religion both have their own "bodies of knowledge." Science's body of knowledge is not universally agreed upon. At the cutting edge of science, there are vigorous debates. Scientists try to come up with tests to show that one hypothesis is better than another. Between scientific fields there is even more disagreement. Cultural anthropologists and behavioral geneticists often have very different scientific views, as do economists and sociologists, even though they might be studying the same phenomena. Although there is a great deal of agreement about what the body of knowledge is, there is disagreement too.

However, the disagreement in science pales in comparison to the divides in the body of knowledge of religion, so much so that it's hard to justify religion in general has having a body of knowledge. Different religious traditions can have vastly different, mutually contradictory views of the world and will sometimes go to war over them (when it comes to resolving intellectual disagreements, I prefer running experiments to violence, but maybe that's just my bias as a scientist). Some religions believe that your behavior in this life will determine what kind of creature you will be reincarnated into. Others believe that your soul will live in some parallel world after death. Yet others believe that your ghost will haunt this world for a time and then just vanish. They can't all be true. I have a friend who majored in religion to find "the common denominator." Those who seek some kind of underlying truth in all world religions will only find similarities at the grossest, most abstract level, such as "there's something more out there." It's probably best to view religion's body of knowledge as specific to each religious tradition. When we speak of "science versus religion," in terms of bodies of knowledge, it makes more sense to speak of science's body of knowledge versus the body of knowledge of some *particular* religion.

Second, just like religion, science is more than a collection of beliefs, and both religion and science have a method for generating those beliefs. Anthropologist Pascal Boyer reports that according to the Fang religion, ghosts exist,[36] and physics claims that electrons exist, but both enterprises had some means of coming up with those beliefs in the first place, and (often) further *justifications* for why someone should believe them to be true today. The Fang and physicists have different ways of knowing. Philosophers would say that these are different epistemologies.

I have heard religious apologists attack science's body of knowledge on the grounds that scientists of the present believe things (and expect us to believe things) that contradict the beliefs of scientists of the past. How can we trust science if its body of knowledge is changing? How do we know that today's scientists are right?

There are two responses to this critique. First, although a given religion might feel eternal and unchanging, it is certainly not. You can read about any religion that has been studied historically and find out how the accepted beliefs have changed over the years. So any critique of science's beliefs changing over the years can be equally applied to any religion.

Second, I would think we would have more justification for worry if science *didn't* change. There are more scientists working today than ever before. There is still so much that we do not understand about this world. And yes, some things we thought we understood will get replaced with better explanations and theories. This is a strength, not a weakness, of science.

If you look at all the scientific ideas that there ever were, you'd probably have to say that many of them got things wrong, or at the very least not completely right. So the probability of some random scientific idea from all of history being correct is not good. But what about *current* scientific ideas? Well, even scientists will admit that much of our understanding of how the world works will likely be overturned in the future by better understandings. We just don't know which ones, nor what they will be replaced with, which makes our understanding of dubious value—the best we can do is to have a general humility. So we continue to believe scientific consensus on a particular issue even though in the back of our minds we know that it might be wrong. This gives some people reason to doubt all of science's body of knowledge.

So is the epistemology of science *better* than the epistemology of religion? Scientific practice has its problems, but even so I think the answer is a definite yes. Scientific ideas might start out on equal footing with religious ones, but the beautiful thing about science is that it requires testing. Many scientific theories make predictions in the world, ideally unexpected ones, that *anybody* can test—and should get the same results. If you fail to get the predicted results, or even if you are the *only one* who gets the predicted results, science will (eventually) weed that idea out. One thing I've learned in being a scientist is that no matter how convinced you are of an idea's rightness, experimentation gives the universe an opportunity to show you that you are completely wrong. No degree of conviction or even careful reasoning will save you. Science has a self-correcting mechanism built into its epistemology that religion lacks. That is why it is a superior epistemology.

To say that science has a superior epistemology is not to say that everything scientists are saying today is right. Far from it. It is to say, however, that what scientists are saying currently is *what is most rational to believe* until they come up with something better.[37]

How do religious beliefs come into existence? The answer is complicated, and I hope that this book has shed some light on how it happens. But to some extent it doesn't matter. Great scientific ideas can come into existence for silly reasons, as in the apocryphal apple that fell on Newton's head. What matters is what comes after. How do religious ideas change? Again, I have tried to describe some ways in this book. But religious beliefs don't change using the same mechanisms as science.

In conclusion, the sciences provide a better epistemology than religions do, and as a result we should put more trust in the scientific body of knowledge than any religious one. If your goal is to *believe things that are closer to the truth,* science is the way to go.

But truth isn't everything. If you're looking to *endorse beliefs that will help hold society together* (as a result of people believing them), well, even science shows that science doesn't fare so well.

*　*　*

Atheists often explain religion as evolutionary cheesecake: we didn't evolve with cheesecake in our evolutionary environment, but we evolved with other foods that have qualities similar to cheesecake (sugar, fat, etc.). Similarly, some believe, people only find religion compelling because it has elements that we evolved to like for other reasons. Is religion just a function of a bunch of other cognitive quirks that spreads like a mental epidemic? Atheists also typically see religion as irrational. But as psychologist Jonathan Haidt persuasively puts it, they are missing the point of religion.[38] For Haidt, "Religion cannot be studied in lone individuals any more than hivishness can be studied in lone bees." He and others have argued that religion is adaptive for societies because it helps to maintain social order. Indeed, religious people give more to charity, but this might be only because they give more to religious charities. Also, people are more likely to think that supernatural agents have "socially strategic" knowledge of when people break rules, supporting the idea that religion is there to foster group cohesion. According to this view, religion is an innovation (either genetic or cultural or some combination of the two) that promotes order and in-group loyalty that help groups compete with other groups. Human beings appear to be the only animals that will sacrifice themselves for groups of others they are not related to.[39]

There are some compelling reasons to believe this. Religious people tend to have more children than the nonreligious. In one study of 200 nineteenth-century American intentional communities (communes), Richard Sosis found that after 20 years, only 6 percent of the secular communes were still functioning, in contrast to the 39 percent of the religious communes. The biggest factor in predicting the success of the commune was the number of costly sacrifices people were required to make (e.g., not drinking alcohol, dress codes). The more sacrifices religious communities required, the longer they lasted. Not so for the secular communes, for which there was no correlation between sacrifices and commune longevity. Why? The theory is that religions make the sacrifices feel sacred.[40] Some things are done, or believed, and are not subjected to the normal cost-benefit analysis that other decisions are, such as picking which brand of almond butter to buy. According to Sosis and Haidt, by making things sacred, religion solves the inevitable problem communities face: how do you get people who are unrelated to not cheat each other? Members of secular communities, on the other hand, who are more likely to see self-sacrificing gestures in a more tit-for-tat way, will not be as likely to blindly help other people in the commune. Without something to make sacrifice sacred, people will think of most decisions in terms of costs and benefits, and societies (at least the size of communes) will fail.

Even if the atheists are correct that religious beliefs are not *rationally* justified, an argument could be made that they are *practically* justified, because it seems that people following religious rules blindly can be good for societal order. As Jonathan Haidt puts it, behaviors that are irrational at an individual level can be thought of as rational at the group level.[41] Believing in religious facts might well be irrational, but it seems that these irrational beliefs were the means by which evolution, cultural or otherwise, got us into functioning societies. Creating a secular society is trying to outsmart evolution and, as such, we can expect it to be damn well difficult. (One strange ray of hope shows that thinking about science *also* makes people exhibit more morally good behavior!) This is not to say that religion is the *only* way to make people behave well in a community.[42] It's just the one that humans have tended to use for at least 10,000 years.

In this book I've discussed religion as though it consists of a set of beliefs and practices that individuals accept or reject, which might be missing the point if the function of religion is to help people get along well in groups. I treat religion as a set of beliefs and practices because no matter what the function of religion is for a society, each individual has to be "sold" on the religion that they are exposed to. The psychological reasons for which they find religion compelling are somewhat independent of the larger societal reasons the religion exists in the first place.

*   *   *

Well, perhaps religion is good for society but is it good or bad for the individual? Should people be rational all the time, or can belief in things not supported by some combination of science and reason sometimes be beneficial?

I realize that I'm assuming here that believing in religion isn't rational, which will probably make my religious readers bristle. But note that I'm talking about *all* religions, not (just) yours. You can't consistently believe in all religions, nor (I imagine) do my religious readers believe that it is rational for a given individual to believe in whatever religion their family happens to have. *Your* religion might very well be perfectly rational, but one can't rationally believe many at the same time, because they are mutually contradictory. So if you're religious, you can read this book as though I'm not talking about your religion, but one of the thousands of others that have existed now and in the history of humanity.

Back to the subject. Is rationality all it's cracked up to be? As much as I adore the idea of being rational all the time, there are some instances when being irrational is actually good for you.

Many of these cases involve beliefs about one's self and one's abilities. For example, people who have a very realistic view of their abilities have "depressive realism," which people with clinical depression tend to have. These people are also less prone to the magical thinking of schizotypals described in chapter 3. Being very skeptical about supernatural events correlates with an inability to experience pleasure. In a study of competitive swimmers, those who were more honest with themselves and didn't deceive themselves tended to perform worse in the competitions.[43] So even though being rational might be better in terms of being *right,* being right isn't all there is to life. It's not always better to be realistic if your goal is happiness or success.

Even belief in the supernatural can sometimes be helpful. The Mayan belief in protecting forest spirits helped the Mayans to protect their forests. A study by psychologist Lysann Damisch found that good luck charms (when you know you have them) increase performance by 50 percent. People with faith are less likely to panic under pressure, and prayer helps people deal with their negative emotions, reduces stress, lowers blood pressure and feels good, due to the release of dopamine.[44]

Religion seems to make people make better health choices: they exercise more, get married, have friends, drink less, and don't smoke. These things lead to a longer life. Mormons, who tend to have community support, which reduces stress, and who don't drink or smoke, tend to live *ten years* longer than the average American, at least in part just due to health. Attending church correlates positively with health (but, interestingly, having religious beliefs does not). On the other hand it has been found that people with HIV who believe that God decides the fate of disease are less likely to take their medicine. What's going on here? It turns out that going to church is good for your health if you are uneducated, but not if you're educated. Church is good for getting you off of drugs and smoking, which both educated and uneducated people benefit from. But it can also have the problematic effect of undermining belief in evidence-based medicine and science, which exacerbates the problems that tend to face the educated, who are more likely to have access to medical diagnoses and treatments.[45] There is an overall healthful effect on the population as a whole simply because uneducated people are more likely to go to church. Church has similar effects on the rich and poor, but that effect ends up helping the poor and hurting the rich because of their different circumstances.

Being orthodox reduces the chances that a people will have their children vaccinated but also reduces anxiety in general. People who believe in past lives have less stress about dying. Religious people tend to have lower activation in the part of the brain called the anterior cingulate cortex (ACC), which is associated with self-regulation and the experience of anxiety. Indeed, studies show that the deeply religious experience less anxiety when committing errors. Even just hearing about religion can have this effect. The ACC is associated with critical thinking in general, so that's the tradeoff. Religious people tend to be happier, but that might be because they have more social interaction, because they believe they have meaning in their lives, or because they are more certain of their beliefs in general. People with weak religious conviction are less happy than agnostics or the strongly religious. Also, the increase in happiness goes away in societies in which religion is not highly valued.[46] In America, for example, it's highly valued.

It is tempting for atheists to make the following argument: belief in any known religion is irrational; therefore anybody who believes in a religion is irrational. There is an unstated premise in this argument, though, and that is that if someone believes in some things that are irrational, then they can be classified as generally irrational people. But this is silly, because all of us believe irrational things at one time or another, and this does not render us forever untrustworthy with thinking in general. The terms *rational* and *irrational* are better suited for particular instances of thinking rather than broad stokes used to describe people.

Further, there is some evidence that people use different kinds of reasoning when thinking about religion than they use when thinking about ordinary things. For example, someone might truly believe that the communion host is the body of Jesus without being willing to put it to any empirical tests, but at the same time work as a biochemist and be very rigorous in his or her laboratory when testing materials. Religions would not have survived if believing in them made people *generally* irrational in their day-to-day lives. As Scott Atran eloquently puts it, "The trick is in knowing how and when to suspend factual belief without countermanding the facts and compromising survival."[47] Just because someone believes in God does not mean they can't be a rigorous, skeptical thinker when it comes to mundane aspects of life, such as figuring out why someone is late to dinner or where they left their umbrella.

It could be that religious believers do not justify religious beliefs according to rationality, but rather emotional satisfaction. Rather than using observations and evidence as religious beliefs' primary justification, Atran holds that they are justified by how they satisfy primary emotional drives such as the desire for justice and order, as well as relief of anxieties.[48] This is not to say that people do not eventually draw upon observations and rationality to *further* justify their beliefs—they certainly do—but to say that religion is primarily justified emotionally.

What about morality? Religious people commonly wonder how atheists could have any morality at all. Indeed, religious people give more to charity. Reminding people about God makes them more charitable, but interestingly this has the same effect on religious believers as atheists, and a similar effect can be produced by reminding people of secular moral institutions, such as the law. Perhaps this is because one of the roles religion has played in human societies is keeping us from being too selfish.

The morality that is presumably from God is not as crystal clear and unambiguous as it might first appear. All Christians are "grocery store" Christians, choosing what from the Bible they believe. There is a lot in the Bible that Christians of our modern era do not accept. Why? It does not sit right with our modern sensibilities. Why don't we burn witches anymore? People have *opinions* of what God thinks, even with the Bible and religious leaders as general guides. What morals Christians take and don't take from the Bible are based on culture and personal preference, and *atheists choose their morals in the same way*. Certainly religion can affect people's morality, but for the most part people get it backward: it's not religion that shapes people's morals, but people's morals that shape religion.

How do we know this is true?

In most places in the world people do not get to choose their religion in the way we think of doing in the Western world. Most people are born into a cultural context in which the tenets of the culture's religion are presented as facts along with other, nonreligious information, such as what plants are safe to eat. But readers of this book might be in a place where there are many religions that one could potentially take up.

Even within a religion, people will often change denominations, perhaps because the first one was too serious, or not serious enough. I have Jewish friends who have chosen to attend a Conservative synagogue and others who attend a Reform synagogue, and they both say that they do it because they are more in agreement with the practices of whichever one they chose. This anecdotal evidence is only suggestive, but is presented as an easily understood example that readers might be familiar with of how the expressed beliefs of a branch of a religion are expected to align with people's personal ideas of the right way to do something.

Similarly, we all probably know people who are part of religious groups, but do not agree with some subset of that group's moral beliefs.

If you are religious, reflect on how your morals might have changed over the course of your life, or perhaps over the course of the lives of people you know. Sometimes a religious teaching can convince someone that something is right or wrong, but you might also have experienced the opposite, where people change their minds about whether something is morally acceptable or not (e.g., using curse words, birth control) and then claim that God agrees with their new moral stance. What's interesting is that in the past they believed that God agreed with their *old* moral stance. What seems to be happening is that people attribute their own moral beliefs to God, rather than trying to change what they think is right or wrong according to some objective judgment about what God thinks.

Indeed, experimental evidence by business researcher Nicholas Epley has shown that this is the case. In fact, when people's beliefs were *manipulated* in the laboratory (the subjects were given persuasive essays to read), their ideas about what God believed changed too! People think God agrees with them even more than they think other people agree with them. The same study showed that when thinking about God's beliefs, your brain's activation looks more like it does when thinking about your own beliefs than it does thinking about other people's beliefs.[49]

Just about all of us have morals. Like most things, people's morality probably has a strong genetic component. The difference is that atheists do not try to *justify* their morals with a religion.

* * *

In the introduction I described the impossibility critique, that understanding things like art and religion is too complicated and varied for science. Scholars in general can be broken down into two groups: lumpers and splitters. The splitters enjoy showing the diversity of the world and delight in finding exceptions to generalizations. Lumpers, on the other hand, like to see things as being the same, to find commonalities, and to classify things. In fact, fitting all scholars into just two groups like lumpers and splitters is a very lumper thing to do.

This book is super lumpy. What I have presented here is not a knock-down set of experiments showing that all things we love are compelling for the same reasons. I have described in general terms a possible research program, a way to look at these phenomena that ties them together. My hope is that compellingness foundations theory will encourage future researchers to explore some of the theory's many testable predictions, and to give nonscientists a new way to examine their lives and the things they find riveting.

What we find compelling is to a great extent preprogrammed into our brains. Just as we inherited earlobes and toenails, we also inherited brain processes that helped our ancestors thrive. Although we are fascinated by the differences between the ways of one society compared to another, successful cultural ideas must conform to these inborn brain processes as they are. In this book I've explored six dimensions that influence what we find compelling in what we experience and shown that these same dimensions have ramifications for what we find compelling in a variety of things, including art, religion, sports, and explanations. Those six dimensions are the social compellingness theory; hopes and fears; the fact that we delight in patterns; our motivation to explore incongruities; psychological biases; and biological influences.

Some of these processes leave us ill-equipped to deal with the ways of our modern world, such as our love for eating sugar. They also leave us ill-equipped to deal with propagandists and advertisers who play to these tendencies deliberately to get what they want from us. History is full of cultural problems that were manufactured by companies selling the solutions. The practical application of compellingness foundations theory is twofold.

First, we can use knowledge of what's compelling to help us make more compelling art. The empirical study of the arts complements the folk wisdom that teachers of the arts have passed down through the ages. It can reveal, first of all, whether or not these teachers have been right. But more than that, it can control variables so that we can know very precisely what is working and why. Although much of art making involves subconscious processes, art training (including books on how to paint or write fiction) includes explicitly articulated advice that students take to heart.

Second, we can use knowledge of what makes ideas compelling to help us make decisions about what to believe (to the extent that we are able). When we encounter ideas, we have our habits and old-brain processes weighing in on the quality of those ideas. Our old brain (to personify it) has opinions and biases and, when it speaks up, our knowledge of its biases can help us evaluate the quality of its opinion. Does my old brain like this idea simply because a beautiful woman is telling it to me, or because it's written beautifully, or because I'm being exposed to an anecdote framed in the form of a compelling story? Or maybe all at once?

Or does the old brain dislike the idea only because it's being spoken by someone who does not speak the language well, or because it is presented as only numbers, without a human angle, or because the idea is boring? None of those things mean it isn't true, even though it can feel less compelling.

The mind is an association maker. Every bit of information it takes in, it feeds into its huge, unconscious pattern-detecting machinery. With this in mind, one should be wary of the kind of input this machinery receives. Just as you would not want to put data into a spreadsheet or computer model that is not representative of the system you are trying to understand, exposing yourself to too much anomalous information will give you a skewed take on our world. If you expose yourself to lots of horrible stories, you will be scared.

Be wary of compelling ideas that play on our biological natures or on the various biases we have in our psyches, whether they are based on evolution or cultural learning. Be wary of compelling ideas that are framed in terms of people and relationships, are easy to understand, present an intriguing puzzle, or play to our hopes and fears. Belief systems are often accepted or rejected not by evidence, but by how they suit our psychological needs.

But all the while, revel in that great feeling that compellingness provides. Love art, love ideas, whether you accept them or not. Let yourself get riveted. It is one of the exquisite joys of being human.

MORE SOURCES OF INTERESTINGNESS

There are several excellent books about religion and its psychological underpinnings. Casual readers will enjoy the following:

Jonathan Haidt, *The Righteous Mind: Why Good People Are Divided By Politics and Religion* (Pantheon Books, 2012).
Jesse Bering, *The Belief Instinct: The Psychology of Souls, Destiny, and the Meaning of Life* (W. W. Norton, 2011).
Pascal Boyer, *Religion Explained: The Evolutionary Origins of Religious Thought* (Basic Books, 2001).

For a detailed description of the psychology of paranormal claims, aliens, and conspiracy theories, I recommend Michael Shermer, *The Believing Brain: From Ghosts and Gods to Politics and Conspiracies: How We Construct Beliefs and Reinforce Them as True* (Times Books, 2011).

For an extended argument that our artistic appreciation is the result of evolutionary adaptations, see Dennis Dutton, *The Art Instinct: Beauty, Pleasure, and Human Evolution* (Bloomsbury Press, 2009). For an argument that we evolved to *create* art, see Ellen Dissanayake, *Homo Aestheticus: Where Art Comes From and Why* (University of Washington Press, 1995).

For a readable introduction to the difference between the old brain and the new, I recommend Keith E. Stanovich, *The Robot's Rebellion* (University of Chicago Press, 2004).

ACKNOWLEDGMENTS

A big thank you goes to my agent, Don Fehr at Trident Media, my editor Elisabeth Dyssegaard, my copyeditor Steven Wagley, Katherine Haigler, and the whole team at Palgrave Macmillan.

This work benefited from my conversations with my wife, Vanessa Davies, and many people who are not my wife, including Michael Shermer, anthropologist Ellen Dissanayake, art scholar Denis Dutton, psychologist Jonathan Haidt, religion scholar Johannes Wolfart, philosopher Heidi Maibom, my photographer and all-around smart guy Daniel Thompson, psychologist Daniel Saunders, film and stage director Montica Pes, microbiologist Alex Gill, novelist Johannah D. Spero, game designer Lou Fasulo, and theater professor Daniel Mroz. Much of the color research was gathered by Jobina Li.

Thank you to my research assistants, who helped me collect the references for this book, including Heather Burch, Jessica Cockbain, RhiAnne Brown, Leah MacQuarrie, Michelle Sappong, Sterling Somers, and Sarah Lacelle.

Thanks to the people who read and commented on drafts of this book: Janet and James Davies, philosophers Peter Slezak and Jeanette Bicknell, computer scientist and novelist Anthony Francis, science podcaster Darren McKee, and especially English professor Donald Beecher.

NOTES

INTRODUCTION

1. Blackmore, S. J. (1996). *In search of the light: Adventures of a parapsychologist*. Prometheus.
2. Limb, C. J., & Braun, A. R. (2008). Neural substrates of spontaneous musical performance: An fMRI study of jazz improvisation. *PLoS ONE, 3*(2), e1769.
3. I should note that nearly all of the citations in this book are from multiauthored papers. So rather than using the cumbersome "Joshua Greene and colleagues" throughout the book, I will use only one of the authors, usually the first; Greene, J. D., Sommerville, R. B., Nystrom, L. E., Darley, J. M., & Cohen, J. D. (2001). An fMRI investigation of emotional engagement in moral judgment. *Science, 293*(5537), 2105–2108; As this book was going to press, I read Greene's new book. In it he claims that the brain area associated with more utilitarian thinking is the DLPFC and the deontological area is the VMPFC. These two areas are both in the frontal lobe, and, in contrast with general tendencies in the brain, the utilitarian thinking area is further forward than the deontological one. It could be an exception to the general frontal/deliberative and back-of-the-brain/hardwired pattern often found in the brain; Greene, J. (2013). *Moral tribes: Emotion, reason, and the gap between us and them*. Penguin.
4. DeCaro, M. S., Thomas, R. D., & Beilock, S. L. (2008). Individual differences in category learning: Sometimes less working memory capacity is better than more. *Cognition, 107*(1), 284–294.
5. Gendler calls this quasi-belief an "alief," a "mental state triggered by ambient environmental factors, generating very real emotional and behavioral responses, but the person experiencing that mental state isn't convinced that the trigger reflects something true." Gendler, T. S. (2008). Alief and belief. *The Journal of Philosophy, 105*(10), 634–663. Cited in: Bering, J. (2011). *The belief instinct: The psychology of souls, destiny, and the meaning of life*. W. W. Norton.
6. Boden, M. T., Berenbaum, H., & Topper, M. (2012). Intuition, affect, and peculiar beliefs. *Personality and Individual Differences, 52*(7), 845–848.
7. Davis, H., & McLeod, S. L. (2003). Why humans value sensational news. *Evolution and Human Behavior, 24*(3), 208–216.
8. Curry, A. (2010a). The mathematics of terrorism. *Discover, 31*(6), 38–43.
9. Terrorism leads to fear: Healy, A. F., Aylward, A. G., Bourne Jr., L. E., & Beer, F. A. (2009). Terrorism after 9/11: Reactions to simulated news reports. *American Journal of Psychology, 122*(2), 153–165; Crime documentaries:

Kort-Butler, L. A., & Sittner Hartshorn, K. J. (2011). Watching the detectives: Crime programming, fear of crime, and attitudes about the criminal justice system. *Sociological Quarterly, 52*(1), 36–55; Appel, M. (2008). Fictional narratives cultivate just-world beliefs. *Journal of Communication, 58*(1), 62–83.

10. Dutton, D. (2009). *The art instinct: Beauty, pleasure, and human evolution* (p. 110). Bloomsbury Press.

11. Marsh, E. F., & Fazio, L. K. (2006). Learning errors from fiction: Difficulties in reducing reliance on fictional stories. *Memory & Cognition, 34*(5), 1140–1149.

12. Byrne, D. (2012). *How music works*. San Francisco, CA: McSweeny's.

13. For a review see: Winner, E. (1982). *Invented worlds: The psychology of the arts* (pp. 66–67). Harvard University Press.

14. Dutton, D. (2009). *The art instinct* (p. 30).

15. McAdams, D. P. (2001). The psychology of life stories. *Review of General Psychology, 5*(2), 100–122.

16. Empathy: Mar, R. M., Oatley, K., Hirsh, J., de la Paz, J., & Peterson, J. B. (2006). Bookworms versus nerds: Exposure to fiction versus non-fiction, divergent associations with social ability, and the stimulation of fictional social worlds. *Journal of Research in Personality, 40*(5), 694–712; Prosocial behavior: Johnson, B. (2012). Religion and philanthropy. Unpublished manuscript, University of North Carolina at Chapel Hill. African Americans: Katz, P. A., & Zalk, S. R. (1978). Modification of children's racial attitudes. *Developmental Psychology, 14*(5), 447–461; No mental state descriptions: Peskin, J., & Wilde Astington, J. (2004). The effects of adding metacognitive language to story texts. *Cognitive Development, 19*(2), 253–273.

17. Marsh, E. F., & Fazio, L. K. (2006). Learning errors from fiction.

18. Boyer, P. (2001). *Religion explained: The evolutionary origins of religious thought*. Basic Books, 132.

19. MacCoby, E. E. (1998). *The two sexes: Growing up apart, coming together (the family and public policy)*. Harvard University Press; Baumeister, R. F., & Sommer, K. L. (1997). What do men want? Gender differences and two spheres of belongingness: Comment on Cross and Madson. *Psychological Bulletin, 122*(1).

20. Gal, D., & Rucker, D. D. (2010). When in doubt, shout!: Paradoxical influences of doubt on proselytizing. *Psychological Science, 21*(11), 1701–1707. This causes an interesting interpretation of the death of Jesus. Early disciples followed Jesus and then he was killed. Perhaps their evangelism is a result of their savior dying.

21. Epstein, S., Pacini, R., Denes-Raj, V., & Heier, H. (1996). Individual differences in intuitive-experiential and analytical-rational thinking styles. *Journal of Personality and Social Psychology, 2*(71), 390–405; Pennycook, G., Cheyne, J. A., Seli, P., Koehler, D. J., & Fugelsang, J. A. (2012). Analytic cognitive style predicts religious and paranormal belief. *Cognition, 123*(3), 335–346.

22. Waller, N. G., Kojelin, B. A., Bouchard, T. J., Lykken, D. T., & Tellegen, A. (1990). Genetic and environmental influences on religious interests, attitudes, and values: A study of twins reared apart and together. *Psychological Science, 1*(3), 138–142. Perhaps even more surprisingly, it's the same for people's political beliefs.

23. Clow, A., & Fredhoi, C. (2006). Normalisation of salivary cortisol levels and self-report stress by a brief lunchtime visit to an art gallery by London city workers. *Journal of Holistic Healthcare, 3*(2), 29–32.

24. Rice, T. W. (2003). Believe it or not: Religious and other paranormal beliefs in the United States. *Journal of the Scientific Study of Religion, 42*(1), 95–106.

25. Moore, S. G. (2012). Some things are better left unsaid: How word of mouth influences the storyteller. *Journal of Consumer Research. 38*(6), 1140-1154.

26. Wilson, T. D., Centerbar, D. B., Kermer, D. A., & Gilbert, D. T. (2005). The pleasures of uncertainty: Prolonging positive moods in ways people do not anticipate. *Journal of Personality and Social Psychology, 88*(1), 5–21.

1: HARDWIRING FOR SOCIALIZING

1. It's not clear how long humans have had hierarchical societies. Current nomadic hunter-gatherer societies tend to be egalitarian. It has been suggested that hierarchy became widespread when people became more sedentary and started using agriculture: Haidt, J. (2012). *The righteous mind: Why good people are divided by politics and religion.* Pantheon Books. Chapter 8: The Conservative Advantage.

2. In fact, according to the social-intelligence hypothesis, the rapid brain growth of our recent evolutionary history is due to this arms race of our abilities to keep up with complex social networks: Dunbar, R. I. M. (1993). Coevolution of neocortical size, group size, and language in humans. *Behavioral and Brain Sciences, 16*(4), 681–735.

3. Agenticity: Shermer, M. (2011). *The believing brain: From ghosts and gods to politics and conspiracies: How we construct beliefs and reinforce them as true.* Times Books; Hypertrophy of social cognition: Boyer, P. (2003). Religious thought and behavior as by-products of brain function. *Trends in Cognitive Science, 7*(3), 119–124; Overactive theory of mind: Bering, J. (2011). *The belief instinct: The psychology of souls, destiny, and the meaning of life.* W.W. Norton & Company; Hypersensitive agency detection: Haidt, J. (2012). *The righteous mind: Why good people are divided by politics and religion.* Pantheon Books. Dennett, D. C. (2006). *Breaking the spell: Religion as a natural phenomenon.* Penguin Books; Anthropomorphism: Guthrie, S. E. (1993). *Faces in the Clouds: A New Theory of Religion.* Oxford University Press.

4. Isola, P., Xiao, J., Torralba, A., & Oliva, A. (2011). What makes an image memorable? In *IEEE Conference on Computer Vision and Pattern Recognition (CVPR),* (pp. 145-152), Colorado Springs, CO. IEEE.

5. Wilkins, D., Schults, B., & Linduff, K. M. (2008). *Art Past, Art Present.* (S. Touborg, Ed.) (6th ed.). Upper Saddle River, New Jersey: Pearson.

6. Dunbar, R. I. M. (1993). Coevolution of neocortical size, group size, and language in humans. *Behavioral and Brain Sciences, 16*(4), 681–735.

7. Trick, L. M. & Pylyshyn, Z. W. (1994). Why are small and large numbers enumerated differently?: A limited-capacity preattentive stage in vision. *Psychological Review, 101*(1), 80–102.

8. McNeil, W. H. (1995). *Keeping together in time: Dance and drill in human history.* Harvard University Press; For an explanation of why Europeans gave it up, see: Haidt, J. (2012). *The Righteous Mind.* Chapter 10: The Hive Switch.

9. Byrne, D. (2012). *How music works.* San Francisco, CA: McSweeny's. Chapter 2: My Life In Performance.

10. Hogan, P. C. (2003). *Cognitive science, literature, and the arts: A guide for humanists.* Routledge.

11. Zwaan, R. A. (1994). Effect of genre expectations on text comprehension. *Journal of Experimental Psychology: Learning, Memory and Cognition, 20*(4), 920–933.

12. Harris, P. L. (2000). *The work of the imagination.* Blackwell.

13. Gabriel, S., & Young, A. (2011). Becoming a vampire without being bitten: The narrative collective assimilation hypothesis. *Psychological Science, 22*(8), 990–994.

14. Cohen, J. (2004). Parasocial break-up from favorite television characters: The role of attachment styles and relationship intensity. *Journal of Social and Personal Relationships, 21*(2), 187–202.

15. Gilbert, D. T. (1991). How mental systems believe. *American Psychologist, 46*(2), 577–609. Gueraud, S., Harmon, M. E., & Peracchi, K. A. (2005). Updating situation models: The memory-based contribution. *Discourse Processes, 39*(2-3), 243–263.

16. Mar, R. A., & Oatley, K. (2008). The function of fiction is the abstraction and simulation of social experience. *Perspectives on Psychological Science, 3*(3), 173.

17. Castelli, F., Frith, C., Happe, F., & Frith, U. (2002). Autism, Asperger syndrome and brain mechanisms for the attribution of mental states to animated shapes. *Brain: A Journal of Neurology, 125*(8), 1839–1849.

18. Of course I mean realistic according to what people *believe* to be consistent with human psychology (sometimes called "folk psychology"), and not necessarily realistic according to people's *actual* psychology. Storytelling conventions can seem realistic not because we see them played out in real life but because we are used to seeing them in the media. One unrealistic behavior I see in films quite often is when people drop what they're holding when sufficiently surprised. This never happens in real life. In fact, in one women's self-defense course my friend took, they had to train women to drop their bags in order to effectively defend themselves, because normally they won't. Another example is the frequent occurrence of Freudian slips during sexual tension (that is, sexual slips of the tongue, such as saying "just a sex" instead of "just a sec"). As unrealistic as these behaviors are, we find them acceptable in television and movies.

19. Legare, C. H., & Gelman, S. A. (2008). Bewitchment, biology, or both: The co-existence of natural and supernatural explanatory frameworks across development. *Cognitive Science, 32*(4), 607–642.

20. Pronin, E., Wegner, D. M., McCarthy, K., & Rodriguez, S. (2006). Everyday magical powers: The role of apparent mental causation in the overestimation of personal influence. *Journal of Personality and Social Psychology, 91*(2), 218–231.

21. Gray, K., & Wegner, D. M. (2010). Blaming God for our pain: Human suffering and the divine mind. *Personality and Social Psychology Review, 14*(1), 7–16.

22. Schizotypals, whom I will deal with in chapter 3, also tend to like science fiction and fantasy, though I don't have an explanation for this.

23. Zunshine, L. (2006). *Why we read fiction: Theory of mind and the novel.* Ohio State University Press.

24. Connellan, J., Baron-Cohen, S., Wheelwright, S., Batki, A., & Ahluwalia, J. (2000). Sex differences in human neonatal social perception. *Infant Behavior and Development, 23,* 113–118.

25. Johnstone, K. (1999). *Impro for storytellers.* Routledge/Theatre Company.

26. If you are skeptical that a game can be a work of art, I highly recommend an excellent online review of the Russian computer game *Pathologic:* Smith, Q. (2008). *Butchering Pathologic.* Retrieved from http://www.rockpaper shotgun.com/2008/04/10/butchering-pathologic-part-1-the-body/ (March 13, 2011). Another great example is the computer game *The Graveyard.*

27. The NPD Group (2009a). *Entertainment trends in America.* Retrieved from http://www.npd.com/lps/Entertainment_Trends2009/ (March 8, 2011); The NPD Group (2009b). *More Americans play video games than go out to the movies.* Retrieved from https://www.npd.com/wps/portal/npd/us/news/press -releases/pr_090520// (February 2, 2014). Thompson, C. (2007). *Halo 3: How Microsoft labs invented a new science of play.* Retrieved from http://www. wired.com/gaming/virtualworlds/magazine/15-09/ff_halo?currentPage=all (October 18, 2010).

28. Bogost, I. (2007). *Persuasive games: The expressive power of videogames* (page 2). Cambridge, MA: MIT Press.

29. Libby, L. K., Shaeffer, E. M., & Eibach, R. P. (2009). Seeing meaning in action: A bidirectional link between visual perspective and action identification level. *Journal of Experimental Psychology: General, 138*(4), 503–516.

30. Tavinor, G. (2005). Videogames and interactive fiction. *Philosophy and Literature, 29*(1), 24–40.

31. Dunbar, R. I. M. (2004). Gossip in evolutionary perspective. *Review of General Psychology, 8*(2), 100–110.

32. Gossip forms a social bond: De Backer, C., Larson, C., & Cosmides, L. (2007). Bonding through gossip? The effect of gossip on levels of cooperation in social dilemma games. Paper presented at the annual meeting of the International Communication Association, San Francisco, CA; Gossip is often correct: Simmons, D. B. (1985). The nature of the organizational grapevine. *Supervisory Management, 30*(11), 39–42; Gossip prevents selfish behavior: Beersma, B., & Van Kleef, G. A. (2011). How the grapevine keeps you in line: Gossip increases contributions to the group. *Social Psychological and Personality Science, 2*(6), 642–649.

33. Kelly, A. E. (1999). Revealing personal secrets. *Current directions in psychology science, 8*(4), 105–109.

34. Boyer, P. (2001). *Religion explained: The evolutionary origins of religious thought* (p. 124). Basic Books.

35. McAndrew, F. T., Bell, E. K., & Garcia, C. M. (2007). Who do we tell, and whom do we tell on? Gossip as a strategy for status enhancement. *Journal of Applied Social Psychology, 37*(7), 1562–1577.

36. Bering, J. (2011). *The belief instinct.*

37. Gambetta, D. (1994). Godfather's gossip. *European Journal of Sociology, 35*(2), 199–223.

38. Purzycki, B. G., Finkel, D. N., Shaver, J., Wales, N., Cohen, A. B., & Sosis, R. (2012). What does god know? Supernatural agents' access to socially strategic and non-strategic information. *Cognitive Science, 36*(5), 846–869.

39. Arena, M. P., & Howe, J. S. (2008). A face can launch a thousand shares—and an 0.80% abnormal return. *Journal of Behavioral Finance, 9*(3), 107–116.

40. Slovic, P. (2007). "If I look at the mass I will never act": Psychic numbing and genocide. *Judgment and Decision Making, 2*(2), 79–95.

41. Hobbs, D. R., & Gallup Jr., G. G. (2011). Songs as a medium for embedded reproductive messages. *Evolutionary Psychology, 9*(3), 390–416.

42. Boyer, P. (2001). *Religion explained* (p. 231).
43. Morris, M. W., Sheldon, O. J., Ames, D. R., & Young, M. J. (2007). Metaphors and the market: Consequences and preconditions of agent and object metaphors in stock market commentary. *Organizational Behavior and Human Decision Processes, 102*(2), 174–192.
44. Kelemen, D. (2004). Are children "intuitive theists"? Reasoning about purpose and design in nature. *Psychological Science.* 15(5), 295–301.
45. Bering, J. (2011). *The belief instinct.*
46. Schoenherr, J., Thomson, R., & Davies, J. (2011). Expanding the space of cognitive science (pp. 1424–1429). Proceedings of the 33rd annual meeting of the cognitive science society, Boston, MA.
47. Novella, S. (2000). *UFOs: The psychocultural hypothesis.* Retrieved from http://www.theness.com/index.php/ufos-the-psychocultural-hypothesis/ (September 13, 2012).
48. Malmstrom, F. V. (2005). Close encounters of the facial kind: Are UFO alien faces an inborn facial recognition template? *Skeptic, 11*(4), 44–47.
49. Harmon, L. D., & Julesz, B. (1973). Masking in visual recognition: Effects of two-dimensional filtered noise. *Science, 180*(4091), 1194–1197.
50. Atkinson, J. (2002). *The developing visual brain.* Oxford University Press.
51. Perina, K. (2004). Cracking the Harvard X-Files. In D. Sobel & J. Cohen (Eds.), *The best American science writing 2004* (pp. 115–123). HarperCollins.
52. Bering, J. (2011). *The belief instinct.*
53. Schjoedt, U., Stødkilde-Jørgensen, H., Geertz, A. W., & Roepstorff, A. (2009). Highly religious participants recruit areas of social cognition in personal prayer. *Social Cognitive and Affective Neuroscience, 4*(2), 199–207.
54. James, W. (1988). *The listening ebony: Moral knowledge, religion, and power among the Uduk.* Clarendon Press.
55. Sanderson, S. K., & Roberts, W. W. (2008). The evolutionary forms of the religious life: A cross-cultural, quantitative analysis. *American Anthropologist, 110*(4) 454–466.
56. Boyer, P. (2001). *Religion explained* (p. 159).
57. Barrett, J. L. (2007). Cognitive science of religion: What is it and why is it? *Religion Compass, 1*(6), 768–786.
58. Caldwell-Harris, C., Fox Murphy, C., Velazquez, T., & McNamara, P. (2011). Religious belief systems of persons with high functioning autism. Annual Meeting of the Cognitive Science Society, Boston, MA.
59. Previc, F. H. (2006). The role of the extrapersonal brain systems in religious activity. *Consciousness and Cognition, 15*(3), 500–539.
60. Norenzayan, A., Gervais, W. M., & Trzesniewski, K. H. (2012). Mentalizing deficits constrain belief in a personal god. *PLoS One, 7*(5), e36880. Paek, E. (2006). Religiosity and perceived emotional intelligence among Christians. *Personality and Individual Differences, 41*(3), 479–490.
61. Discussed in: Atran, S. (2002). *In gods we trust: The evolutionary landscape of religion.* Oxford University Press. Boyer, P. (2001). *Religion Explained* (pp. 145–147).
62. Waytz, A., Morewedge, C. K., Epley, N., Monteleone, G., Gao, J.-H., & Cacioppo, J. T. (2010). Making sense by making sentient: Effectance motivation increases anthropomorphism. *Journal of Personality and Social Psychology, 99*(3), 410-435.
63. Harris, L. T., & Fiske, S. T. (2006). Dehumanizing the lowest of the low: Neuroimaging responses to extreme out-groups. *Psychological Science, 17*(10), 847–853.

64. Boyer, P. (2001). *Religion Explained* (pp. 292–283).

2: WIZARD'S FIRST RULE

1. Goodkind, T. (1994). *Wizard's first rule*. Tor Books.
2. Nasrallah, M., Carmel, D., & Lavie, N. (2009). Murder, she wrote: Enhanced sensitivity to negative word valence. *Emotion, 9*(5), 609–618; Cimpian, A., Brandone, A. C., & Gelman, S. A. (2010). Generic statements require little evidence for acceptance but have powerful implications. *Cognitive Science, 34*(8), 1452–1482.
3. As reported in: Dutton, D. (2009). *The art instinct: Beauty, pleasure, and human evolution (p.* 18). Bloomsbury Press.
4. Cook, M., & Mineka, S. (1990). Selective associations in the observational conditioning of fear in rhesus monkeys. *Journal of Experimental Psychology: Animal Behavior Processes, 16*(4), 372–389; Mineka, S., & Cook, M. (1988). Social learning and the acquisition of snake fear in monkeys. In T. R. Zentall & B. G. Galef (Eds.), *Social learning: Psychological and biological perspectives* (pp. 51–74). Lawrence Erlbaum Associates.
5. People are also more likely to form a food aversion from subsequent nausea than from subsequent unpleasant noise. In this case, they have a built-in expectation that certain foods might make them feel sick, but not that certain foods might produce sounds they find unpleasant. This ends up affecting their beliefs about cause and effect in the world. Thanks to Anthony Francis for this insight. For information about the Baldwin effect, see: Baldwin, M. J. (1896). A new factor in evolution. *The American Natualist, 30*(354), 441–451.
6. Dodd, M. D., Balzer, A., Jacobs, C. M., Gruszczynski, M. W., Smith, K. B., & Hibbing, J. R. (2012). The political left rolls with the good and the political right confronts the bad: Connecting physiology and cognition to preferences. *Philosophical Transactions of the Royal Society B: Biological Sciences, 367*(1589), 640–649.
7. Nial, P., & McGregor, I. (2009). Conservative shift among liberals and conservatives following 9/11/01. *Social Justice Research, 22*(2-3), 231–240.
8. Munro, G. D. (2010). The scientific impotence excuse: Discounting belief-threatening scientific abstracts. *Journal of Applied Social Psychology, 40*(3), 579–600.
9. Kuhn, D., Weinstock, M., & Flaton, R. (1994). How well do jurors reason? Competence dimensions of individual variation in a juror reasoning task. *Psychological Science, 5*(5), 289–296.
10. The classic experimental example of the congruence bias is the Wason card task. In it, participants are asked to turn over cards to see if a given rule is being followed (for example, every card with an even number on one side must have a vowel on the other). Unless the rule is in a domain in which the participants are familiar (e.g., drinking laws), they tend not to turn cards over that would falsify the rule. In other words, they seek information congruent with their belief, rather than information that would refute it. Tut tut. For the Wason card task, see: Wason, P. C. (1960). On the failure to eliminate hypotheses in a conceptual task. *Quarterly Journal of Experimental Psychology, 12*, 129–140.
11. They might indeed be on the same scale, differing only in valence. See: Atran, S. (2002). *In gods we trust: The evolutionary landscape of religion*. Oxford University Press. page 182.
12. Wypijewski, J. (1998). *Painting by numbers: Komar and Melamid's scientific guide to art*. University of California Press.

13. For a discussion, see chapter 1 of: Dutton, D. (2009). *The art instinct*.

14. Balling, J. D., & Falk, J. H. (1982). Development of visual preference for natural environment. *Environment and Behavior, 14*(1), 5–28. Elizabeth Lyons, however, failed to replicate these findings: Lyons, E. (1983). Demographic correlates of landscape preference. *Environment and Behavior, 15,* 487–511.

15. These findings are from: Orians, G. H., & Heerwagen, J. H. (1992). Evolved responses to landscapes. In J. H. Barkow, L. Cosmides, & J. Tooby (Eds.), *The adapted mind: Evolutionary psychology and the generation of culture.* Oxford University Press. And from: Appleton, J. (1975). *The experience of landscape.* Wiley. Cultural influences, of course, can override these instincts. The British are known for surrounding their properties with tall, opaque hedges. Presumably they value privacy more than their ability to see their surroundings from their homes.

16. Sporrle, M., & Stich, J. (2010). Sleeping in safe places: An experimental investigation of human sleeping place preferences from an evolutionary perspective. *Evolutionary Psychology, 8*(3), 405–419.

17. Curry, A. (2010b). Where the wild things are. *Discover, 31*(2), 58–65.

18. Lyons, E. (1983). Demographic correlates of landscape preference; Synek, E., & Grammer, K. (1998). *Evolutionary aesthetics: Visual complexity and the development of human landscape preferences.* Retrieved from http://evolution.anthro.univie.ac.at/institutes/urbanethology/projects/urbanisation/landscapes/indexland.html (May 26, 2007).

19. Nisbet, E. K., & Zelenski, J. M. (2011). Underestimating nearby nature: Affective forecasting errors obscure the happy path to sustainability. *Psychological Science, 22*(9), 1101–1106.

20. Pettijohn, T. F., II , & Sacco Jr., D. F. (2009). Tough times, meaningful music, mature performers: Popular billboard songs and performer preferences across social and economic conditions in the USA. *Psychology of Music, 37*(2), 155–179.

21. Knobloch-Westerwick, S., Gong, Y., Hagner, H., & Kerbeykian, L. (2012). Tragedy viewers count their blessings: Feeling low on fiction leads to feeling high on life. *Communication Research 40,* 747—766.

22. There is evidence to suggest that there is a social reason for horror movies too. Couples leaving thrillers were observed to be especially likely to be touching each other in some way, such as holding hands: Wiseman, R. (2011). *59 seconds: Change your life in under a minute.* Anchor Books. Perhaps fear drives the couples closer together because they feel they need to bond together to resist a common threat. Men who watch horror films with women enjoy them more if the women appear more frightened, and women enjoy them more if the man appears unmoved: Zillmann, D., Weaver, J. B., Mundorf, N., & Aust, C. F. (1986). Effects of an opposite-gender companion's affect to horror on distress, delight, and attraction. *Journal of Personality and Social Psychology, 51*(3), 586–594.

23. Revonsuo, A. (2000). The reinterpretation of dreams: An evolutionary hypothesis of the function of dreaming. *Behavioral and Brain Sciences, 23*(6), 877–901.

24. Yet another theory holds that dreaming is one part of the mind trying to make sense of junk from some other part of the mind. I don't see this as a competing theory, however, because the *way* the junk gets interpreted still needs an explanation, and it could be that the junk gets interpreted as threat-simulation narratives. In other words, both theories could be right.

25. Gackenbach, J., & Kuruvilla, B. (2008). The relationship between video game play and threat simulation dreams. *Dreaming, 18*(4), 236–256. Computer

gaming appears to have other benefits. The competence of a surgeon is well-predicted by the number of hours clocked playing them. Playing violent computer games can increase pain tolerance, as well as aggression: Stephens, R., & Allsop, C. (2012). Effect of manipulated state aggression on pain tolerance. *Psychological Reports: Disability and Trauma, 111,* 311–321.

26. I will note that finding research on the Web for this kind of happy ending is difficult because of the abundance of the other kind. McGranahan, D. V., & Wayne, I. (1948). German and American traits reflected in popular drama. *Human Relations, 1*(4), 429–455.

27. For the distinction between obsessive and pleasurable activities see: Marsolais, J. (2003). Les passions de l'âme: On obsessive and harmonious passion. *Personality and Social Psychology, 85*(4), 756–767.

28. Winstanley, C. A., Cocker, P. J., & Rogers, R. D. (2011). Dopamine modulates reward expectancy during performance of a slot machine task in rats: Evidence for a "near-miss" effect. *Neuropsychopharmacology, 36*(5), 913–925.

29. Fox Tree, J. E., & Weldon, M. S. (2007). Retelling urban legends. *American Journal of Psychology, 120,* 459–476.

30. Boustany, N. (2005). Wealthy Muslim nations do little to stop spread of polio. *Washington Post,* August 17, A09.

31. Mooney, C. (2009). Why does the vaccine/autism controversy live on? *Discover, 30*(6), 58–65.

32. Norenzayan, A., & Hansen, I. G. (2006). Belief in supernatural agents in the face of death. *Personality and Social Psychology, 32*(2), 174–187; Jackson, C. J., & Francis, L. J. (2004). Are interactions in Gray's reinforcement sensitivity theory proximal or distal in the prediction of religiosity: A test of the joint subsystems hypothesis. *Personality and Individual Differences, 36*(5), 1197–1209.

33. On loneliness, see: Gebauer, J. E., & Maio, G. R. (2012). The need to belong can motivate belief in God. *The Journal of Personality, 80*(2), 465–501. On experienced terrorism, see: Mohmand, M. G. K., Ibrahim, H., Khan, I. S., Akram, U., & Hasnain, F. (2011). On the constant threat of terrorism: Stress levels and coping strategies amongst university students of Karachi. *The Journal of the Pakistan Medical Association, 61*(4), 410–414. On insecurity, see: Bartkowski, J. P., Xu, X., & Garcia, G. E. (2011). Religion and infant mortality in the United States: A preliminary study of denominational variations. *Religions, 3*(2), 264–276. On being anxious, see: McGregor, I., Nash, K., & Prentice, M. (2010). Reactive approach motivation (RAM) for religion. *Journal of Personality and Social Psychology, 99*(1), 148–161.

34. Dysfunctional: Paul, G. (2009). The chronic dependence of popular religiosity upon dysfunctional psychosociological conditions. *Evolutionary Psychology, 7*(3), 398–441; Lower standard of living: Rahman, T., Mittelhammer, R. C., & Wandschneider, P. R. (2011). Measuring quality of life across countries: A multiple indicators and multiple causes approach. *Journal of Socio-Economics, 40*(1), 43–52; Inequality: Ruiter, S., & van Tubergen, F. (2009). Religious attendance in cross-national perspective: A multilevel analysis of 60 countries. *American Journal of Sociology, 115*(3), 863–895; Less trust: Berggren, N., & Bjornskov, C. (2011). Is the importance of religion in daily life related to social trust? Cross-country and cross-state comparisons. *Journal of Economic Behavior & Organization, 80*(3), 459–480; Less democratic: Buhlmann, M., Merkel, W., & Muller, B. (2011). Denmark, Finland and Belgium have best democracies. http://www.mediadesk.uzh.ch /articles/2011/schweizer-demokratie_en.html. According to Gallup World-View, although the United States appears to be quite religious compared to

other industrialized countries, only 65 percent of citizens polled reported that religion was "important to daily life." There are over 100 countries more religious than the United States on this measure, with thirteen countries rating 98 percent or higher, including Bangladesh, Indonesia, and Malawi.

35. Rees, T. (2009a). Atheist nations are more peaceful. http://epiphenom.fieldof science.com/2009/06/atheist-nations-are-more-peaceful.html. Rees, T. J. (2009b). Is personal insecurity a cause of cross-national differences in the intensity of religious belief? *Journal of Religion and Society, 11,* 1–24. Immerzeel, T., & van Tubergen, F. (2011). Religion as reassurance? Testing the insecurity theory in 26 European countries. *European Sociological Review, OnlineFirst,* 1–14. Atran, S. (2002). *In gods we trust* (p. 182).

36. Atran, S. (2002). *In gods we trust. (p. 75).*

37. Boyer, P. (2001). *Religion explained: The evolutionary origins of religious thought* (p. 20). Basic Books.

38. Epley, N., Akalis, S., Waytz, A., & Cacioppo, J. T. (2008). Creating social connection through inferential reproduction: Loneliness and perceived agency in gadgets, gods, and greyhounds. *Psychological Science, 19*(2), 114–120. Hutson, M. (2012). The unbearable uncanniness of being. *Psychology Today, 45*(4), 50–59. Kay, A. C., Moscovitch, D. A., & Laurin, K. (2010). Randomness, attributions of arousal, and belief in God. *Psychological Science, 21*(2), 216–218.

39. Bering, J. (2011). *The belief instinct: The psychology of souls, destiny, and the meaning of life.* W.W. Norton. Chapter 4: Curiously Immortal.

40. Our minds could potentially survive the death of our bodies if those minds were replicated in another substrate, such as some new, manufactured brain, or perhaps if the information in our minds was uploaded to a computer and could function as a program. What I'm saying, more specifically, is that it is absurd to think that our minds could survive brain death if the information in that brain is not copied to some other functioning system.

41. Treisman, D. (2011). The geography of fear. *National Bureau of Economic Research.* Working Paper No. 16838: http://www.nber.org/papers/w16838.

42. Ellis, L., Wahab, E. A., & Ratnasingan, M. (2012). Religiosity and fear of death: A three-nation comparison. *Mental Health, Religion, & Culture, OnlineFirst,* 1–21.

43. Boyer, P. (2001). *Religion explained* (pp. 19, 280).

44. Gal, D., & Rucker, D. D. (2010). When in doubt, shout!: Paradoxical influences of doubt on proselytizing. *Psychological Science, 21*(11), 1701–1707.

45. Pettazzoni, R. (1955). On the attributes of God. *Numen, 2*(1-2), 1–27.

46. Roes, F. L., & Raymond, M. (2003). Belief in moralizing gods. *Evolution and Human Behavior, 24*(2), 126–135.

47. Epley, N., Converse, B. A., Delbosc, A., Monteleone, G. A., & Cacioppo, J. T. (2009). Believer's estimates of God's beliefs are more egocentric than estimates of other people's beliefs. *Proceedings of the National Academy of Sciences of the United States of America, 106*(51), 21533–21538.

48. Boyer, P. (2001). *Religion explained (p. 139).*

49. Abalakina-Paap, M., Stephan, W. G., Craig, T., & Gregory, W. L. (1999). Beliefs in conspiracies. *Political Psychology. Special Issue: Political socialization, 20*(3), 637–647.

3: THE THRILL OF DISCOVERING PATTERNS

1. Gopnik, A. (2000). Explanation as orgasm and the drive for causal understanding: The evolution, function and phenomenology of the theory-forma

tion system. In F. Keil & R. Wilson (Eds.), *Cognition and explanation* (pp. 299–323). MIT Press.

2. Shermer, M. (2011). *The believing brain: From ghosts and gods to politics and conspiracies: How we construct beliefs and reinforce them as true* (p. 5). Times Books.

3. Schilling, M. F. (1990). The longest run of heads. *College Mathematics Journal, 21*(3), 196–207.

4. Horry, R., Wright, D. B., & Tredoux, C. G. (2010). Recognition and context memory for faces from own and other ethnic groups: A remember-know investigation. *Memory & Cognition, 38*(2), 134–141. This might be part of why we have what is called the "out-group homogeneity bias," which makes us think that our in-groups are more variable than groups we are not in; see: Quattrone, G. A., & Jones, E. E. (1980). The perception of variability within in-groups and out-groups: Implications for the law of small numbers. *Personality and Social Psychology, 38*(1), 141–152.

5. Schmidhuber, J. (2007). Simple algorithmic principles of discovery, subjective beauty, selective attention, curiosity & creativity. *Lecture Notes in Computer Science, 4755,* 26–38.

6. Hekkert, P., & van Wieringen, P. C. W. (1996). The impact of level of expertise on the evaluation of original and altered versions of post-impressionistic paintings. *Acta Psychologica, 94*(2), 112–131.

7. Carbon, C.-C., & Leder, H. (2005). The repeated evaluation technique (RET). A method to capture dynamic effects of innovativeness and attractiveness. *Applied Cognitive Psychology, 19*(5), 587–601.

8. Critchley, M. (1970). *The dyslexic child.* William Heinemann Medical Books. Lachmann, T., & van Leeuwen, C. (2007). Paradoxical enhancement of letter recognition in developmental dyslexia. *Developmental Neuropsychology, 31*(1), 61–77.

9. Dehaene, S., & Cohen, L. (2007). Cultural recycling of cortical maps. *Neuron, 56*(2), 384–398.

10. Oppenheimer, D. M. (2008). The secret life of fluency. *Trends in Cognitive Science, 12*(6), 237–241.

11. Lev-Ari, S., & Keysar, B. (2010). Why don't we believe non-native speakers? The influence of accent on credibility. *Journal of Experimental Social Psychology, 46*(6), 1093–1096.

12. Adee, S. (2012). Tricksy type: How fonts can mess with your mind. *New Scientist, 216*(2896–2897), 68–69.

13. Simonton, D. K. (1994). *Greatness: Who makes history and why* (pp. 234–236). Guilford Press.

14. Begg, I. M., Anas, A., & Farinacci, S. (1992). Dissociation of processes in belief: Source recollection, statement familiarity, and the illusion of truth. *Journal of Experimental Psychology: General, 121,* 446–458.

15. Changizi, M. A., Zhang, Q., Hao, Y., & Shimojo, S. (2006). The structures of letters and symbols throughout human history are selected to match those found in objects in natural scenes. *American Naturalist, 167*(5), 117–139.

16. These contrasting sentences are attributed to novelist E. M. Forster: Forster, E. M. (1927). *Aspects of the novel.* Harcourt, Brace.

17. Johnstone, K. (1999). *Impro for storytellers.* Routledge/Theatre Company.

18. One of the problems with interactive fiction is that with open-ended plots, leave-behinds are difficult to create. Ideally, a narrative is a unified whole, with the ending and the beginning composed so that they mutually

reinforceeachother.See:http://complicationsensue.blogspot.com/2011/02/how
-branching-stories-fail.html.

19. Moretti, F. (2009). Style, INC. reflections on seven thousand titles (British
novels, 1740–1850). *Critical Inquiry, 36*(1), 134–158.

20. Huron, D. (2006). *Sweet anticipation: Music and the psychology of expectation.* MIT Press.

21. Bolinger, D. (1986). *Intonation and its parts: Melody in spoken English.* Stan-
ford University Press.

22. McGlone, M. S., & Tofighbakhsh, J. (2000). Birds of a feather flock conjointly
(?): Rhyme as reason in aphorisms. *Psychological Science, 11*(5), 424–428.

23. Scrabble has made me a fan of using the proper names of letters to refer to
them, e.g., *ay, bee, cee, dee,* etc.

24. The popular conception of schizophrenia is a confusion with dissociative
identity (multiple-personality) disorder. Schizophrenia is actually characterized by a disorganization of thought and a reduction in emotional responsive-
ness. Common symptoms include auditory hallucinations and delusions. John
Nash, the Princeton mathematician portrayed in the film *A Beautiful Mind,*
had schizophrenia.

25. Kapur, S. (2003). Psychosis as a state of aberrant salience: A framework link-
ing biology, phenomenology, and pharmacology in schizophrenia. *American
Journal of Psychiatry, 160*(1), 13–23.

26. Brugger, P., Landis, T., & Regard, M. (1990). A "sheep-goat effect" in repetition avoidance: Extra-sensory perception as an effect of subjective probability? *British Journal of Psychology, 81*(4), 455–468.

27. Krummenacher, P., Mohr, C., Haker, H., & Brugger, P. (2009). Dopamine,
paranormal belief, and the detection of meaningful stimuli. *Journal of Cognitive Neuroscience, 22*(8), 1670–1681.

28. Atran, S. (2002). *In gods we trust: The evolutionary landscape of religion.* (p.
188 and 194). Oxford University Press.

29. Previc, F. H. (2006). The role of the extrapersonal brain systems in religious
activity. *Consciousness and Cognition, 15*(3) (p. 514), 500–539. Pechey, R.,
& Halligan, P. (2011). The prevalence of delusion-like beliefs relative to socio-
cultural beliefs in the general population. *Psychopathology, 44*(2), 106–115.

30. Sapolsky, R. M. (1997). *The trouble with testosterone: And other essays on
the biology of the human predicament,* chapter "Circling the blanket for
God" (pp. 241–288). Scribner.

31. Previc, F. H. (2006). The role of the extrapersonal brain systems in religious
activity. (p. 525); Radin, P. (1987). *The trickster: A study in American Indian
mythology.* Schocken. The popular media often mentions that there are particular genes for diseases. This is a convenient shorthand, but it's important to
understand that nature would have selected out genes that caused serious illnesses very often. The more cumbersome, accurate way to put it is that certain
genes, when expressed too little or too much, or in a particular environment,
are a causal factor in etiology. Normal gene expression is often helpful: one
gene protects us from malaria but when overexpressed (when it is too productive) can cause sickle-cell anemia. Another protects us from tuberculosis but
when overexpressed can give us Tay-Sachs disease. The same gene implicated
in cystic fibrosis might protect us from cholera (there's some evidence for this).
Likewise, Radin offers us an explanation for how the "genes for schizophrenia" might have been replicated in human history through the social acceptance of schizotypals as shamans.

32. Horrobin, D. F. (1998). Schizophrenia: The illness that made us human. *Medical Hypotheses, 50*(4), 269–288.

33. *The encyclopedia of mental disorders* (2011), Schizotypal personality disorder. Retrieved from http://www.minddisorders.com/ (January 27, 2011).

34. Rogers, P., Davis, T., & Fisk, J. (2009). Paranormal belief and susceptibility to the conjunction fallacy. *Applied Cognitive Psychology, 23*(4), 524–542.

35. I am not referring to the religion known as Wicca when I use the terms witches and witchcraft. What I mean by witchcraft is the anthropological definition: the casting of spells by people to hurt other people. Wicca is a particular European religion that is more positive, and is certainly not representative of witchcraft, as defined by anthropologists, worldwide. See: Boyer, P. (2001). *Religion explained: The evolutionary origins of religious thought* (pp. 193–194). Basic Books.

36. Swami, V., Coles, R., Stieger, S., Pietschnig, J., Furnham, A., Rehim, S., & Voracek, M. (2011). Conspiracist ideation in Britain and Austria: Evidence of a monological belief system and associations between individual psychological differences and real-world and fictitious conspiracy theories. *British Journal of Psychology, 102*(3), 443–463.

37. Shermer, M. (2011). *The believing brain* (p. 126).

38. Wood, M. J., Douglas, K. M., & Sutton, R. M. (2012). Dead and alive: Beliefs in contradictory conspiracy theories. *Social Psychological and Personality Science, OnlineFirst*. Retrieved February 7, 2014 from: http://spp.sagepub.com /content/3/6/767.short.

39. Wisneski, D., Lytle, B., & Skitka, L. (2009). Gut reactions: Moral conviction, religiosity, and trust in authority. *Psychological Science, 20*(9), 1059–1063.

40. Finkelhor, D., Williams, L., Burns, N., & Kalinowski, M. (1988). Executive summary—sexual abuse in day care: A national study. University of New Hampshire, Family Research Laboratory. Retrieved from https://www.ncjrs .gov/pdffiles1/Digitization/113095NCJRS.pdf (July 17, 2013).

41. Lipsett, A. (2008). Under-fives have almost no male role models. *The Guardian*, August 7. Interestingly, though, one study failed to find that parents were prejudiced against male day care workers: Wessell, M. E. (1986). Sex-role orientation and attitudes towards male day care workers. Master's thesis, Humboldt State University. It could be that men's fears are mostly unfounded.

42. Bering, J. (2011). *The belief instinct: The psychology of souls, destiny, and the meaning of life*. W.W. Norton. Chapter 3: Signs, signs, everywhere signs.

43. Xie, J., Sreenivasan, S., Korniss, G., Zhang, W., Lim, C., & Szymanski, B. K. (2011). Social consensus through the influence of committed minorities. *Physical Review E: Statistical, Nonlinear, and Soft Matter Physics, 84*(1), 011130 1–8.

44. Stanovich lists four excellent rules a person (which he calls the "vehicle," because it is the vehicle that genes use to propagate themselves) should follow when accepting memes (ideas): "1. Avoid installing memes that are harmful to the vehicle physically. 2. Regarding memes that are beliefs, seek to install only memes that are true—that is, that reflect the way the world actually is. 3. Regarding memes that are desires, seek to install only memes that do not preclude other memeplexes becoming installed in the future. 4. Avoid memes that resist evaluation"; Stanovich, K. E. (2004). *The robot's rebellion* (p. 185). University of Chicago Press.

45. Legare, C. H., & Souza, A. L. (2012). Evaluating ritual efficacy: Evidence from the supernatural. *Cognition, 124*(1), 1–15; Norton, M. I., & Gino, F.

(forthcoming). Rituals alleviate grieving for loved ones, lovers, and lotteries. *Journal of Experimental Psychology: General.* When people do rituals before high anxiety tasks, they end up calmer and more confident, resulting in better performance on that task; see: Brooks, A., Schroeder, J., Risen, J., Gino, F., Galinsky, A., & Norton, M. (2013). Don't stop believing: Coping with anxiety through rituals. Unpublished manuscript.

46. Whitehouse, H. (2000). *Arguments and icons: Divergent modes of religiosity* (p. 46). Oxford University Press.

47. Boyer, P. (2001). *Religion explained* (p. 191).

48. Bartlett, F. (1932). *Remembering.* Cambridge, UK: Cambridge University Press. Cited in Atran, S. (2002). *In gods we trust.* (p. 258)

49. Boyer, P. (2001). *Religion explained* (pp. 163–164).

50. People remember minimally counterintuitive ideas better than mundane or merely counterfactual ones in certain contexts. Technically, they degrade less in memory even if they are not recalled as well; see: Atran, S. (2002). *In gods we trust.* (p. 102–107).

51. This terminology is from J. L. Barrett (see below) and: Boyer, P. (2001). *Religion explained.* Technically, minimally counterintuitive concepts are members of some "ontological category" (person, animal, plant, artifact, natural/nonliving) with some deviation from the normal state of affairs in some "intuitive knowledge domain" (psychology, biology, physics). These categories form a 15-cell matrix, each cell of which can be filled with ideas that are pretty good candidates for religious inclusion. For example, if you take an ANIMAL and violate something about its PHYSICS, you might get the idea of a "bird that can turn invisible." If you take a PERSON who violates the PSYCHOLOGY domain, you might have a woman who can read minds. See a chart with more examples in: Barrett, J. L. (2000). Exploring the natural foundations of religion. *Trends in Cognitive Sciences,* 4(1), 29—34.

52. Kelly, M. H., & Keil, F. C. (1985). The more things change . . . : Metamorphoses and conceptual structure. *Cognitive Science,* 9(4), 403–416.

53. Although I believe this is the first book to seriously consider that religion and art are compelling for similar reasons, Scott Atran hints at it in his book: Atran, S. (2002). *In gods we trust.*

54. Trefil, J. (1997). *101 things you don't know about science and no one else does either.* Cassell Illustrated.

55. Bryson, B. (2003). *A short history of nearly everything* (p. 214). Transworld Publishers.

56. Woese, C. R. (1998). Default taxonomy: Ernst Mayr's view of the microbial world. *Proceedings of the National Academy of Sciences of the United States of America,* 95(19), 11043–11046.

57. Mayr, E. (1998). Two empires or three? *Proceedings of the National Academy of Sciences of the United States of America,* 95(17), 9720–9723.

4: INCONGRUITY

1. Hoppe, C., & Stojanovic, J. (2008). High-aptitude minds. *Scientific American Mind, 19,* 60–67.

2. Bering, J. (2011). *The belief instinct: The psychology of souls, destiny, and the meaning of life.* W.W. Norton. Chapter 6: God as adaptive illusion.

3. Brockman, J. (Ed.). (2007). *What is your dangerous idea? Today's leading thinkers on the unthinkable* (p. 252). Harper Perennial.

4. Atran, S. (2002). *In gods we trust: The evolutionary landscape of religion.* Oxford University Press. (p. 36).

5. Richerson, P. J., & Boyd, R. (2008). *Not by genes alone: How culture transformed human evolution.* University of Chicago Press. Chapter 4: Culture is an adaptation.

6. The cerebral cortex looks much the same everywhere, in contrast with our mid- and hind-brain areas, which are more modular and function specific. This would be expected for the part of our brain theorized to be the general-purpose learner. See: Hawkins, J., & Blakeslee, S. (2004). *On intelligence.* Times Books.

7. Shaw, P., Greenstein, D., Lerch, J., Clasen, R., Lenroot, N., A., G., & Giedd, J. (2006). Intellectual ability and cortical development in children and adolescents. *Nature, 440*(7084), 676–679.

8. Gopnik, A. (2000). Explanation as orgasm and the drive for causal understanding: The evolution, function and phenomenology of the theory-formation system. In F. Keil & R. Wilson (Eds.), *Cognition and explanation,* (pp. 299–323). MIT Press.

9. Smets, G. (1973). *Aesthetic judgment and arousal: An experimental contribution to psycho-Aesthetics.* Leuven University Press; Munsinger, H., & Kessen, W. (1964). Uncertainty, structure, and preference. *Psychological Monographs: General & Applied, 78*(9, whole no. 586), 24.

10. Day, H. (1967). Evaluations of subjective complexity, pleasingness, and interestingness for a series of random polygons varying in complexity. *Perception & Psychophysics, 2*(7), 281–286.

11. Thompson, C. (2007). *Halo 3: How Microsoft labs invented a new science of play.* Retrieved from http://www.wired.com/gaming/virtualworlds/magazine/15–09/ff_halo?currentPage=all (October 18, 2010).

12. Hekkert, P., Snelders, D., & van Wieringen, P. C. W. (2003). "Most advanced, yet acceptable": Typicality and novelty as joint predictors of aesthetic preference in industrial design. *British Journal of Psychology, 94*(1), 111–124.

13. Phillips, F., Norman, J. F., & Beers, A. M. (2010). Fechner's aesthetics revisited. *Seeing and Perceiving, 23*(3), 263–271.

14. Martindale, C., Moore, K., & Borkum, J. (1990). Aesthetic preference: Anomalous findings for Berlyne's psychobiological theory. *American Journal of Psychology, 103*(1), 53–80.

15. Bhatara, A., Tirovolas, A. K., Duan, L. M., Levy, B., & Levitin, D. J. (2011). Perception of emotional expression in musical performance. *Journal of Experimental Psychology, 37*(3), 921–934.

16. Moseman, A. (2010). Hot science: The best new science culture. *Discover, 31,* 30–36.

17. Leonard, A. (2007). William Gibson. *Rolling Stone, 1039,* 162.

18. Taylor, G. T. (1992). "The cognitive instrument in the service of revolutionary change": Sergei Eisenstein, Annette Michelson, and the avant-garde's scholarly aspiration. *Cinema Journal, 31*(4), 42–59.

19. Feist, G. J., & Brady, T. R. (2004). Openness to experience, non-conformity, and the preference for abstract art. *Empirical Studies of the Arts, 22*(1), 77–89.

20. De Vries, M., Holland, R. W., Chenier, T., Starr, M. J., & Winkielman, P. (2010). Happiness cools the warm glow of familiarity: Psychophysiological evidence that mood modulates the familiarity-affect link. *Psychological Science, 21*(3), 321–328.

21. This would be because our stress and exhaustion drain our mental energy that we would need to appreciate incongruity. For my theory behind this, see: Davies, J., & Fortney, M. (2012). The menton theory of engagement and boredom. In Langley, P. (Ed.), *Proceedings of the First Annual Conference on Advances on Cognitive Systems* (pp. 131–143).

22. Schüll, N. D. (2012). *Addiction by design: Machine gambling in Las Vegas.* Princeton University Press.

23. Denworth, L. (2013). Brain-changing games. *Scientific American Mind, 23*(6), 28–35. Dye, M. W. G., Green, C. S., & Bavelier, D. (2009). Increasing speed of processing with action video games. *Current Directions in Psychological Science, 18*(6), 321–326. Jackson, L. A., Witt, E. A., Games, A. I., Fitzgerald, H. E., von Eye, A., & Zhao, Y. (2012). Information technology use and creativity: Findings from the children and technology project. *Computers in Human Behavior, 28*(2), 370–376. Maurer, D. (2012). Lessons about visual plasticity from adults treated for congenital contracts. *Annual Meeting of the American Association for the Advancement of Science (AAAS-2012), 6638.* McFarlane, A., Sparrowhawk, A., & Heald, Y. (2002). *Report on the educational use of games. An exploration by TEEM of the contribution which games can make to the education process.* Department for Education and Skills. Technical report.

24. Pandelaere, M., Millet, K., & Van den Bergh, B. (2010). Madonna or Don McLean? The effect of order of exposure on relative liking. *Journal of Consumer Psychology, 20*(4), 442–451.

25. I should mention that this is not a rigorously done scientific study, but was more informal: Capps, R. (2009). *The good enough revolution: When cheap and simple is just fine.* Retrieved from http://www.wired.com/gadgets/miscellaneous/magazine/17-09/ff_goodenough?currentPage=all (January 28, 2011).

26. Krishna Rao, H. P. (1923). *The psychology of music.* Guluvias Printing Works.

27. Bowling, D. L., Sundararajan, J., Han, S., & Purves, D. (2012). Expression of emotion in Eastern and Western music mirrors vocalization. *PloS ONE, 7*(3), e31942.

28. Byrne, D. (2012). *How music works.* San Francisco, CA: McSweeny's.

29. Blanchard-Fields, F., Coon, R. C., & Mathews, R. C. (1986). Inferencing and television: A developmental study. *Journal of Youth and Adolescence, 15*(6), 453–459.

30. Miura, N., Sugiura, M., Takahashi, M., Sassa, Y., Miyamoto, A., Horie, K., Sato, S., Nakamura, K., & Kawashima, R. (2010). Effect of motion smoothness on brain activity while observing a dance: An fMRI study using a humanoid robot. *Social Neuroscience, 5*(1), 40–58; Christensen, J. F. & Calvo-Merino, B. (2013). Dance as a subject for empirical aesthetics. *Psychology of Aesthetics, Creativity, and the Arts, 7*(1), 76–88.

31. Byrne, D. (2012). *How music works.* Chapter 10: Harmonia mundi.

32. Lasher, M. D., Carroll, J. M., & Bever, T. G. (1983). The cognitive basis of aesthetic experience. *Leonardo,* 16,196–199.

33. Bastian, B., Jetten, J., & Fasoli, F. (2011). Cleansing the soul by hurting the flesh: The guilt-reducing effect of pain. *Psychological Science, 22*(3), 334–335.

34. Billings, J., & Sherman, P. W. (1998). Antimicrobial functions of spices: Why some like it hot. *Quarterly Review of Biology, 73*(1), 3–49.

35. Rozin, P. (1996). Towards a psychology of food and eating: From motivation to module to model to marker, morality, meaning, and metaphor. *Current Directions in Psychological Science, 5*(1), 18–24.

36. Dewaele, J.-M. (2004). The emotional force of swearwords and taboo words in the speech of multilinguals. *Journal of Multilingual and Multicultural Development, 25*(2), 204–222.

37. Katz, B. F. (1993). A neural resolution of the incongruity-resolution and incongruity theories of humor. *Connection Science, 5*(1), 59–75.

38. Aronson, E., & Mills, J. (1959). The effect of severity of initiation on liking for a group. *Journal of Abnormal and Social Psychology, 59*(2), 177–181.

39. Some of these ideas were published in: Davies, J. (2012). Academic obfuscations: The psychological attraction of postmodern nonsense. *Skeptic, 17*(4), 44–47.

40. Mishra, H., Mishra, A., & Shiv, B. (2011). In praise of vagueness: Malleability of vague information as a performance booster. *Psychological Science, 22*(6), 733–738.

41. McNamara, D. S., & Healy, A. F. (2000). A procedural explanation of the generation effect for simple and difficult multiplication problems and answers. *Journal of Memory and Language, 43*(4), 652–679.

42. Yang, J., Bachrati, C., Hickson, I., & Brown, G. (2012). BLM and RMI1 alleviate RPA inhibition of topol II_ decatenase activity. *PLoS ONE, 7*(7), e41208.

43. Boden, M. (2003). *The creative mind: Myths and mechanisms.* Routledge.

44. Richards, I. A. (1929). *Practical criticism.* Harcourt Brace. Richards, I. A. (1960). *Variant readings and misreading* (pp. 241–252). Technology Press of Massachusetts Institute of Technology; Rosenblatt, L. M. (1978). *The reader, the text, the poem: The transactional theory of the literary work.* Southern Illinois University Press.

45. Whitehouse, H. (2000). *Arguments and icons: Divergent modes of religiosity* (p. 116). Oxford University Press.

46. Aronson, E., & Mills, J. (1959). The effect of severity of initiation on liking for a group. *Journal of Abnormal and Social Psychology, 59* (2), 177–81.

47. Campbell, J., & Moyers, B. (1988). *The power of myth* (p. 174). Doubleday.

48. Christian fundamentalism, and its focus on literal interpretation of the Bible, is much younger than most people realize. It was a movement that arose in Britain and America in the late 1800s and early 1900s. It is by no means any kind of return to a traditional way of being Christian. It is a departure from how Christian doctrine was interpreted for over a thousand years. See: Sandeen, E. R. (1970). *The roots of fundamentalism: British and American millenarianism, 1800–1930.* University of Chicago Press.

49. Atran, S. (2002). *In gods we trust.* (p. 274).

50. Dennett, D. C. (2006). *Breaking the spell: Religion as a natural phenomenon.* Penguin Books.

51. Mishra, H., Mishra, A., & Shiv, B. (2011). In praise of vagueness: Malleability of vague information as a performance booster. *Psychological Science, 22*(6), 733–738.

52. See http://www.telegraph.co.uk/news/newstopics/howaboutthat/ufo/9653499/UFO-enthusiasts-admit-the-truth-may-not-be-out-there-after-all.html.

53. Bering, J. (2011). *The belief instinct.*

54. As an atheist, I don't see the point in capitalizing *he* when referring to God.

55. Cited in: Boyer, P. (2001). *Religion explained: The evolutionary origins of religious thought (p. 245).* Basic Books.

56. Barth, F. (1975). *Ritual knowledge among the Baktaman of New Guinea.* Yale University Press.

57. Gervais, W. M., & Norenzayan, A. (2012). Analytic thinking promotes religious disbelief. *Science, 336*, 493–496.

58. Boyer, P. (2001). *Religion explained.*

59. Proulx, T., & Heine, S. J. (2009). Connections from Kafka: Exposure to meaning threats improves implicit learning of an artificial grammar. *Psychological Science, 20*(9), 1125–1131.

60. Peskin, J., & Wilde Astington, J. (2004). The effects of adding metacognitive language to story texts. *Cognitive Development, 19*(2), 253–273. Although this is not true for young children who can understand mental states of story characters but need them explicitly mentioned. See: Winner, E. (1982). *Invented worlds (p. 299).*

61. As of 2008 the official database of the Star Wars Universe has over 30,000 entries, and at the time of this writing a full-time employee at Lucasfilm spent about three quarters of his day updating it: Baker, C. (2008). *Meet Leland Chee, the Star Wars franchise continuity cop.* Retrieved from http://www .wired.com/entertainment/hollywood/magazine/16-09/ff_starwarscano n?currentPage=1 (October 2, 2010).

5: OUR BIOLOGICAL NATURE

1. Barrow, J. D. (1995). *The artful universe: The cosmic source of human creativity.* Oxford University Press.

2. Montero, B. (2006). Proprioception as an aesthetic sense. *Journal of Aesthetics and Art Criticism, 64*(2), 231–242.

3. Cross, E. S., Hamilton, A. F. d. C., & Grafton, S. T. (2006). Building a motor simulation de novo: Observation of dance by dancers. *NeuroImage, 31*(3), 1257–1267. Zarorre, R. J., Chen, J. L., & Penhune, V. B. (2007). When the brain plays music: Auditory-motor interactions in music perception and production. *Nature Reviews Neuroscience, 8*(7), 547–558.

4. Koch, S., Holland, R. W., & Knippenberg, A. (2008). Regulating cognitive control through approach-avoidance motor actions. *Cognition, 109*(1), 133–142.

5. The exceptions are interesting. Musicians can often appreciate a piece of sheet music by "playing" the music in their auditory imagination. Experienced actors and directors can read a screenplay or a script and flesh out the performance in their heads. When I was involved with the VisionQuest Theater Company, I read a play by Craig Lucas called *Blue Window* and thought it was terrible. We produced the show and watching it performed was an incredible experience. It was a clear indication to me that I did not yet have a director's vision when reading a play.

6. Lakoff, G., & Johnson, M. (1980). *Metaphors we live by.* University of Chicago Press.

7. Topolinski, S. (2011). I 5683 you: Dialing phone numbers on cell phones activates key-concordant concepts. *Psychological Science, 22*(3), 355–360.

8. Zhong, C.-B., & Leonardelli, G. J. (2008). Cold and lonely: Does social exclusion literally feel cold? *Psychological Science, 19*(9), 838–842.

9. Asch, S. (1955). On the use of metaphor in the description of persons. In Werner, H. (Ed.), *On expressive language: Papers presented at the Clark University conference on expressive language behavior* (pp. 29–38). Clark University Press.

10. Nonreligious tendencies: Brandt, M. J. & Reyna, C. (2011). The chain of being: a hierarchy of morality. *Perspectives on Psychological Science, 6*(5). 428–446; Escalator: Sanna, L. J., Chang, E. C., Miceli, P. M., & Lundberg, K. B. (2011). Rising up to higher virtues: Experiencing elevated physical height

uplifts prosocial actions. *Journal of Experimental Social Psychology, 47*(2), 472–476.

11.   Tolaas, J. (1991). Notes on the origin of some spatialization metaphors. *Metaphor and Symbolic Activity, 6*(3), 203–218.

12.   Previc, F. H. (2006). The role of the extrapersonal brain systems in religious activity. *Consciousness and Cognition, 15*(3), 500–539. Tracy, J. L., & Matsumoto, D. (2008). The spontaneous expression of pride and shame: Evidence for biologically innate nonverbal displays. *Proceedings of the National Academy of Sciences, 105*(33), 11655–11660.

13.   Gray, H. M., Gray, K., & Wegner, D. M. (2007). Dimensions of mind perception. *Science, 315*(5812), 619.

14.   Weak as smell is, training can make it stronger. Scientists have shown that blindfolded and earmuffed people can track a winding chocolate-coated string across a lawn. They end up zigzagging and sniffing a whole lot, just like dogs do. See: Porter, J., Craven, B., Khan, R. M., Chang, S.-J., Kang, I., Judkewicz, B., Volpe, J., Settles, G., & Sobel, N. (2007). Mechanisms of scent-tracking in humans. *Nature Neuroscience, 10*(1), 27–29.

15.   Lee, K., Kim, H., & Vohs, K. D. (2011). Stereotype threat in the marketplace: Consumer anxiety and purchase intentions. *Consumer Research, 38*(2), 343–357. McKeough, T. (2005, September). Down with perfume. *The Walrus*. Retrieved from http://walrusmagazine.com/articles/2005.09-style-making-perfume/ (March 14, 2012).

16.   Kurzweil, R. (2005). *The singularity is near*. Penguin Books.

17.   Barrow, J. D. (1995). *The artful universe* (p. 222).

18.   Magnus, M. (2001). *What's in a word? Studies in phonosemantics*. PhD thesis, Norwegian University of Science and Technology; Emotional connotations: Whissell, C. (1999). Phonosymbolism and the emotional nature of sounds: Evidence of the preferential use of particular phonemes in texts of differing emotional tone. *Perceptual and Motor Skills, 89*, 19–48; Shape connotations: Westbury, C. (2005). Implicit sound symbolism in lexical access: Evidence from an interference task. *Brain and Language, 93*(1), 10–19; *kiki* and *bouba*: Ramachandran, V., & Hubbard, E. M. (2001). Synaesthesia: A window into perception, thought and language. *Journal of Consciousness Studies, 8*(2), 3–34; *oo* is sweet: Simner, J., Cuskley, C., & Kirby, S. (2010). What sound does that taste? Cross-modal mappings across gustation and audition. *Perception, 39*(4), 553–569; Guessing antonyms: Namy, L. L., Nygaard, L. C., Clepper, L., & Rasmussen, S. (2009). Sensitivity to cross-linguistic sound symbolism. *Manuscript in preparation;* Nasal area: Philps, D. (2011). Reconsidering phonaesthemes: Submorphemic invariance in English "sn- words." *Lingua, 121*(6), 1121–1137.

19.   Sizes of things: Ohala, J. J. (1997). Sound symbolism. In *Proceedings of the 4th Seoul International Conference on Linguistics [SICOL]* (pp. 98–103), Seoul, South Korea. Mouth size: Winner, E. (1982). *Invented worlds: The psychology of the arts* (p. 250). Harvard University Press.

20.   Bar, M., & Neta, M. (2007). Visual elements of subjective preference modulate amygdala activation. *Neuropsychologia, 45*(10), 2191–2200. Peña, M., Mehler, J., & Nespor, M. (2011). The role of audiovisual processing in early conceptual development. *Psychological Science, 22*(11), 1419–1421.

21.   Abel, G. A., & Glinert, L. H. (2008). Chemotherapy as language: Sound symbolism in cancer medication names. *Social Science & Medicine, 66*(8), 1863–1869.

22. Fernald, A., & Morikawa, H. (1993). Common themes and cultural variations in Japanese and American mothers' speech to infants. *Phonetica, 57,* 242–254.

23. Dissanayake, E. (1995). *Homo aestheticus: Where art comes from and why* (p. 82). University of Washington Press. Reichel-Dolmatoff, G. (1978). *Beyond the Milky Way: Hallucinatory imagery of the Tukano Indian.* UCLA Latin American Center Publications.

24. Dutton, D. (2009). *The art instinct: Beauty, pleasure, and human evolution* (p. 35). Bloomsbury Press.

25. Schachner, A., Brady, T. F., Pepperberg, I. M., & Hauser, M. D. (2009). Spontaneous motor entrainment to music in multiple vocal mimicking species. *Current Biology, 19*(10), 831–836.

26. Fritz, T., Jentschke, S., Gosselin, N., Sammler, D., Peretz, I., Turner, R., Friederici, A. D., & Koelsch, S. (2009). Universal recognition of three basic emotions in music. *Current Biology, 19*(7), 573–576. Bresin, R., & Friberg, A. (2000). Emotional coloring of computer-controlled music performance. *Computer Music Journal, 24*(4), 44–63.

27. Bowling, D. L., Sundararajan, J., Han, S., & Purves, D. (2012). Expression of emotion in Eastern and Western music mirrors vocalization. *PloS ONE, 7*(3), e31942.

28. Davidson, J. (2006). Measure for measure. *The New Yorker, August 21,* 60–69.

29. DeBodt, E. (2010). Prosodic cues to emotion: Perceptual and acoustic analyses. Master's thesis, Carleton University.

30. Mobbs, D., & Watt, C. (2011). There is nothing paranormal about near-death experiences: How neuroscience can explain seeing bright lights, meeting the dead, or being convinced you are one of them. *Trends in Cognitive Science, 15*(10), 447–449.

31. Shermer, M. (2011). *The believing brain: From ghosts and gods to politics and conspiracies: How we Construct beliefs and reinforce them as true* (p. 104). Times Books. Page 104. Geiger, J. (2009). *The third man factor: The secret to survival in extreme environments* (3rd ed.). Penguin.

32. Atran, S. (2002). *In gods we trust: The evolutionary landscape of religion.* Oxford University Press. Chapter 4: Counterintuitive worlds.

33. Osgood-Hynes, D. (n.d.) Thinking bad thoughts. *MGH/McLean OCD Institute.* Mash, E. J., & Wolfe, D. A. (2005). *Abnormal child psychology (3rd ed).* Thomson-Wadsworth. Elkin, G. D. (1999). *Introduction to clinical psychiatry.* McGraw-Hill Medical Publishing.

34. Previc, F. H. (2006). The role of the extrapersonal brain systems in religious activity. *Consciousness and Cognition, 15*(3), 500–539. Aardema, F., & O'Connor, K. (2007). The menace within: Obsessions and the self. *International Journal of Cognitive Therapy, 21,* 182–197.

35. Sinclair Stevenson, M. (1920). *The rites of the twice-born.* Humphrey Milford and Oxford University Press.

36. Ghazali, al- (1106/1953). *The faith and practice of al-Ghazali.* G. Allen and Unwin.

37. Pyysiainen, I. (2012). Cognitive science of religion: State-of-the-art. *Journal for the Cognitive Science of Religion.* 1(1), 5–28.

38. Gonsalvez, C. J., Hains, A. R., & Stoyles, G. (2010). Relationship between religion and obsessive phenomena. *Australian Journal of Psychology, 62*(2), 93–102.

39. Nemeroff, C. J., Brinkman, A., & Woodward, C. K. (1994). Magical contagion and AIDS risk perception in a college population. *AIDS Education and Prevention, 6*(3):249–265. Nemeroff, C., & Rozin, P. (1994). The contagion concept in adult thinking in the United States: Transmission of germs and interpersonal influence. *Ethos, 22*(2), 158–186.

40. Boyer, P. (2001). *Religion explained: The evolutionary origins of religious thought* (p. 134). Basic Books.

41. Griffiths, R. R., Richards, W. A., McCann, U., & Jesse, R. (2006). Psilocybin can occasion mystical experiences having substantial and sustained personal meaning and spiritual significance. *Psychopharmacology, 187*(3), 268–283.

42. Granqvista, P., Fredriksona, M., Ungea, P., Hagenfeldta, A., Valindb, S., Larhammarc, D., & Larssond, M. (2005). Sensed presence and mystical experiences are predicted by suggestibility, not by the application of transcranial weak complex magnetic fields. *Neuroscience Letters, 379*(1), 1–6. Booth, J. N., Koren, S. A., & Persinger, M. A. (2005). Increased feelings of the sensed presence and increased geomagnetic activity at the time of the experience during exposures to transcerebral weak complex magnetic fields. *International Journal of Neuroscience, 115*(7), 1053–1079.

43. Donovan, J. M. (1996). Multiple personality, hypnosis, and possession trance. In R. van Quekelberghe & D. Eigner (Eds.), *Yearbook of cross-cultural medicine and psychotherapy 1994/Jahrbuch fur transkulturelle Medizin und Psychotherapie 1994* (pp. 99–112). VWB press.

44. Wolford, G., Miller, M. B., & Gazzaniga, M. (2000). The left hemisphere's role in hypothesis formation. *Journal of Neuroscience, 20*(6), RC64.

45. Braithwaite, J. J., Samson, D., Apperly, I., Broglia, E., & Hulleman, J. (2011). Cognitive correlates of the spontaneous out-of-body experience (OBE) in the psychologically normal population: Evidence for an increased role of temporal-lobe instability, body-distortion processing, and impairments in own-body transformations. *Cortex, 47*(7), 839–853.

46. Malinowski, B. (Ed.). (1948). *Magic, science and religion and other essays.* Free Press. Gmelch, G. (2000). Baseball magic (revised edition of "Superstition and ritual in American baseball" from *Elysian Fields Quarterly,*11(3), 25–36). McGraw-Hill/Dushkin.

47. Shermer, M. (2011). *The believing brain* (p. 78).

48. Brugger, P., & Mohr, C. (2008). The paranormal mind: How the study of anomalous experiences and beliefs may inform cognitive neuroscience. *Cortex, 44,* 1291–1298. Brugger, P., Gamma, A., Muri, R., Schafer, M., & Taylor, K. I. (1993). Functional hemispheric asymmetry and belief in ESP: Towards a "neuropsychology of belief." *Perceptual and Motor Skills, 77,* 1299–1308.

49. Moore, D. W. (2005). *Three in four Americans believe in paranormal.* Retrieved from http://www.gallup.com/poll/16915/Three-Four-Americans-Believe-Paranormal.aspx (March 24, 2011). Lyons, L. (2005). *Paranormal beliefs come (super)naturally to some.* Retrieved from http://www.gallup.com/poll/19558/Paranormal-Beliefs-Come-SuperNaturally-Some.aspx (March 24, 2011).

50. Ong, J.-R., Hou, S.-W., Shu, H.-T., Chen, H.-T., & Chong, C.-F. (2005). Diagnostic pitfall: Carbon monoxide poisoning mimicking hyperventilation syndrome. *American Journal of Emergency Medicine, 23*(7), 903–904.

6: OUR PSYCHOLOGICAL BIASES

1. The EEA, or environment of evolutionary adaptation, which is generally thought to be a set of preindustrial hunter-gatherer societies, each with around 150 people, during the Pleistocene period.

2. Glassner, B. (1999). *The culture of fear: Why Americans are afraid of the wrong things*. Basic Books.

3. Kort-Butler, L. A., & Sittner Hartshorn, K. J. (2011). Watching the detectives: Crime programming, fear of crime, and attitudes about the criminal justice system. *Sociological Quarterly, 52*(1), 36–55.

4. Glassner, B. (1999). *The culture of fear*.

5. Berlin, B., & Kay, P. (1996). *Basic color terms: Their universality and evolution*. University of California Press.

6. Barrow, J. D. (1995). *The artful universe: The cosmic source of human creativity*. Oxford University Press.

7. Meier, B. P., Robinson, M. D., & Clore, G. L. (2004). Why good guys wear white: Automatic inferences about stimulus valence based on brightness. *Psychological Science, 15*(2), 82–87.

8. Lechner, A., Simonoff, J. S., & Harrington, L. (2011). Color-emotion associations in the pharmaceutical industry: Understanding universal and local themes. *Color Research & Application, 37*(1), 59–71.

9. Guilford, J. P. (1934). The affective value of color as a function of hue, tint, and chroma. *Journal of Experimental Psychology, 17*(3), 342–370. Winner, E. (1982). *Invented worlds: The psychology of the arts (p. 109)*. Harvard University Press.

10. Zhong, C.-B., Bohns, V. K., & Gino, F. (2010). Good lamps are the best police: Darkness increases dishonesty and self-interested behavior. *Psychological Science, 21*(3), 311–314.

11. Sherman, G. D., & Clore, G. L. (2009). The color of sin: White and black are perceptual symbols of moral purity and pollution. *Psychological Science, 21*(8), 1019–1025. Herbert, W. (2009). The color of sin. *Scientific American Mind,* Nov/Dec, 70–71.

12. Nemeroff, C. & Rozin, P. (1994). The contagion concept in adult thinking in the United States: Transmission of germs and interpersonal influence. *Ethos, 22*(2) (p. 169), 158–186.

13. Huibers, M. J., de Graaf, L., Peeters, F. P., & Arntz, A. (2010). Does the weather make us sad? Meteorological determinants of mood and depression in the general population. *Psychiatry Research, 180*(2-3), 143–146.

14. Bubl, E., Kern, E., Ebert, D., Bach, M., & Tebartz van Elst, L. (2010). Seeing gray when feeling blue? Depression can be measured in the eye of the diseased. *Biological Psychiatry, 68*(2), 205–208.

15. Infant preference: Bornstein, M. H. (1975). Qualities of color vision in infancy. *Journal of Experimental Child Psychology, 19*(3), 401–419; Recovery: Barrow, J. D. (1995). *The artful universe;* Emotional reaction: Robertson, S. A. (1996). *Contemporary ergonomics 1996*. CRC Press.

16. Innate meaning: Pryke, S. R. (2009). Is red an innate or learned signal of aggression and intimidation? *Animal Behavior, 78*(2), 393–398; Seeing red when angry: Fetterman, A. K., Robinson, M. D., Gordon, R. D., & Elliot, A. J. (2011). Anger as seeing red: Perceptual sources of evidence. *Social Psychology and Personality Science, 2*(3), 311–316; Mandrills and color: Setchell, J. M., & Wickings, E. J. (2005). Dominance, status signals and coloration in

male mandrills. *Ethology, 111*(1), 25–50; Red circle most dominant: Little, A. C., & Hill, R. A. (2007). Attribution to red suggests special role in dominance signaling. *Journal of Evolutionary Psychology, 5*(1–4), 161–168; Goalie vision: Greenlees, I., Leyland, A., Thelwell, R., & Filby, W. (2008). Soccer penalty takers' uniform color and pre-penalty kick gaze affect the impressions formed of them by opposing goalkeepers. *Journal of Sports Sciences, 26*(6), 569–576; Wearing red and winning: Attrill, M. J., Gresty, K. A., Hill, R. A., & Barton, R. A. (2008). Red shirt color is associated with long-term team success in English football. *Journal of Sports Sciences, 26*(6), 577–582; Opponent behavior: Hill, R. A., & Barton, R. A. (2005). Red enhances human performance in contests: Signals biologically attributed to red coloration in males may operate in the arena of combat sports. *Nature, 435,* 293; Referees and red: Hagemann, N., Strauss, B., & Leibing, J. (2008). When the referee sees red . . . *Psychological Science, 19*(8), 769–711.

17.   Women wearing red: Elliot, A. J. & Niesta, D. (2008). Romantic red: Red enhances men's attraction to women. *Journal of Personality and Social Psychology, 95*(5), 1150–1164. Niesta Kayser, D., Elliot, A. J. & Feltman, R. (2010). Red and romantic behavior in men viewing women. *European Journal of Social Psychology, 40*(6), 901–908; Higher status: Elliot, A. J., Niesta Kayser, D., Greitemeyer, T., Lichtenfeld, S., Gramzow, R. H., Maier, M. A., & Liu, H. (2010). Red, rank, and romance in women viewing men. *Journal of Experimental Psychology: General, 139*(3), 399–417; Excitement: Wexner, L. B. (1954). The degree to which colors (hues) are associated with mood-tones. *Journal of Applied Psychology, 38*(6), 432–435; Heat: Sivik, L. (1997). Color systems for cognitive research. In C. L. Hardin & L. Maffi (Eds.), *Color Categories in Thought and Language* (pp. 163–192). Cambridge University Press.

18.   IQ: Maier, M. A., Elliot, A. J., & Lichtenfeld, S. (2008). Mediation of the negative effects of red on intellectual performance. *Personality and Social Psychology Bulletin, 34*(11), 1530–1540; Red Room: Stone, N. J. (2001). Designing effective study environments. *Journal of Environmental Psychology, 21*(2), 179–190; Creativity: Mehta, R., & Zhu, R. J. (2009). Blue or red? Exploring the effect of color on cognitive task performance. *Science, 323*(5918), 1226–1229.

19.   More creative: Elliot, A. J., Maier, M. A., Moller, A. C., Friedman, R., & Meinhardt, J. (2007). Color and psychological functioning: The effect of red on performance attainment. *Journal of Experimental Psychology: General, 136*(1), 154–168; Vegetation: Ulrich, R. S., & Society of American Florists (2003). Impact of flowers and plants on workplace productivity. *SAF;* Crime reduction: Kuo, F. E., & Sullivan, W. C. (2001). Environment and crime in the inner city: Does vegetation reduce crime? *Environment & Behavior, 33*(3), 343–367.

20.   Evidence for pink effects on crime: Schauss, A. G. (1979). Tranquilizing effect of color reduces aggressive behavior and potential violence. *Journal of Orthomolecular Psychiatry, 8*(4), 218–221; Evidence against: Wise, B. K. & Wise, J. A. (1988). The human factors of color in environmental design: A critical review. Technical Report NASA-CR-177498, N89-15532, University of Washington.

21.   Förster, J., Epstude, K., & Özelsel, A. (2009). Why love has wings and sex has not: How reminders of love and sex influence creative and analytic thinking. *Personality and Social Psychology Bulletin, 35*(11), 1479–1491.

22. Langlois, J. H., Roggman, L. A., Casey, R. J., Ritter, J. M., Rieser-Danner, L. A., & Jenkins, V. Y. (1987). Infant preferences for attractive faces: Rudiments of a stereotype? *Developmental Psychology, 23*(3), 363–369.

23. Maner, J. K., DeWall, C. N., & Gailliot, M. T. (2008). Selective attention to signs of success: Social dominance and early stage interpersonal perception. *Personality and Social Psychology Bulletin, 34*(4), 488–501; Dunn, M. J., & Searle, R. (2010). Effect of manipulated prestige-car ownership on both sex attractiveness ratings. *British Journal of Psychology, 101*(1), 69–80.

24. Gay porn categories: Ogas, O., & Gaddam, S. (2011). *A billion wicked thoughts: What the world's largest experiment reveals about human desire.* Dutton Adult; Content of romance stories: Barrett, D. (2010). *Supernormal stimuli: How primal urges overran their evolutionary purpose.* W. W. Norton.

25. Shackelford, T. K., Schmitt, D. P., & Buss, D. M. (2005). Universal dimensions of human mate preferences. *Personality and Individual Differences, 39*(2), 447–458.

26. Eastwick, P. W., & Finkel, E. J. (2008). Speed-dating: A powerful and flexible paradigm for studying romantic relationship initiation. In S. Sprecher, A. Wenzel, & J. Harvey (Eds.), *Handbook of relationship initiation* (pp. 217–234). Guilford.

27. Karremans, J. C., Verwijmeren, T., Pronk, T. M., & Reitsma, M. (2009). Interacting with women can impair men's cognitive functioning. *Journal of Experimental Social Psychology, 45*(4), 1041–1044.

28. Genetic fitness: Symons, D. (1979). *Evolution of human sexuality.* Oxford University Press; Perceived health: Thornhill, R., & Gangestad, S. W. (1993). Human facial beauty: Averageness, symmetry, and parasite resistance. *Human Nature, 4*(3), 237–269; Actual health: Rhodes, G., Zebrowitz, L. A., Clark, A., Kalick, S. M., Hightower, A., & McKay, R. (2001). Do facial averageness and symmetry signal health? *Evolution and Human Behavior, 22,* 31–46.

29. Weeden, J., & Sabini, J. (2005). 2005. *Psychological Bulletin, 131*(5), 635–653.

30. Perceived: Berry, D. S., & Brownlow, S. (1989). Were the physiognomists right?: Personality correlates of facial babyishness. *Personality and Social Psychology Bulletin, 15*(2), 266–279; Actual: Mueller, U., & Mazur, A. (1997). Facial dominance in Homo sapiens as honest signaling of male quality. *Behavioral Ecology, 8*(5), 569–579; Divorce, infidelity, violence: Booth, A., & Dabbs, J. M., Jr., (1993). Testosterone and men's marriages. *Social Forces, 72,* 463–477.

31. Folstad, I., & Karter, A. J. (1992). Parasites, bright males, and the immunocompetence handicap. *American Naturalist, 139*(3), 603–622.

32. It appears that sexual dimorphism in a species (when males look different from females) often appears when there is competition for mates. See: Richerson, P. J., & Boyd, R. (2008). *Not by genes alone: How culture transformed human evolution.* University of Chicago Press. Chapter 4: Culture is an adaptation. See also the Wikipedia page on sexual dimorphism.

33. Kruger, D. J. (2006). Male facial masculinity influences attributions of personality and reproductive strategy. *Personal Relationships, 13,* 451–463.

34. Conditional mating strategy: Kruger, D. J., & Fitzgerald, C. J. (2011). Reproductive strategies and relationship preferences associated with prestigious and dominant men. *Personality and Individual Differences, 50*(3), 365–360; Cuckold rates: Bellis, M. A., H. K., Hughes, S., & Ashton, J. R. (2005). Measuring paternal discrepancy and its public health consequences. *Journal of*

Epidemiology & Community Health, 59(9), 749–754; Sperm donation: Kruger, D. J. (2006). Male facial masculinity influences attributions of personality and reproductive strategy. *Personal Relationships, 13, 451–463.*

35. Tallness and dominance: Schwartz, B., Tesser, A., & Powell, E. (1982). Dominance cues in nonverbal behavior. *Social Psychology Quarterly, 45*(2), 114–120; Submission in animals: Haidt, J. (2012). *The righteous mind: Why good people are divided by politics and religion.* (p. 142) Pantheon Books. See also the Wikipedia page on sexual dimorphism for more information.

36. Top of computer screen: Meier, B. P., & Dionne, S. (2009). Downright sexy: Verticality, implicit power, and perceived physical attractiveness. *Social Cognition, 27*(6), 883–892; Men prefer low-SES women: Greitemeyer, T. (2007). What do men and women want in a partner? Are educated partners always more desirable? *Journal of Experimental Social Psychology, 43, 180–194.*

37. Although Richard Dawkins popularized the theory in his book *The selfish gene,* Hamilton created what he called "nepotistic altruism" as an alternative theory to group selection.

38. Diamond, J. (1992). *The third chimpanzee: The evolution and future of the human animal* (pp. 101–103). Harper Perennial.

39. Ibid. (p. 108).

40. 0.70 waist-to-hip ratio: Singh, D. (2002). Female mate value at a glance: Relationships of waist-to-hip ratio to health, fecundity, and attractiveness. *Neuroendocrinology Letters, 23*(4), 81–91; Predicts fecundity: Jasienska, G., Ziomkiewicz, A., Ellison, P. T., Lipson, S. F., & Thune, I. (2004). Large breasts and narrow waists indicate high reproductive potential in women. *Proceedings of the Royal Society B: Biological Sciences, 271,* 1213–1217; Predicts Intelligence: Lassek, W. D., & Gaulin, S. J. C. (2008). Waist-hip ratio and cognitive ability: Is gluteofemoral fat a privileged store of neurodevelopment resources? *Evolution and Human Behavior, 29*(1), 26–34.

41. India and China: Singh, D., Renn, P., & Singh, A. (2007). Did the perils of abdominal obesity affect depiction of feminine beauty in the sixteenth to eighteenth century British literature? Exploring the health and beauty link. *Proceedings of the Royal Society B, 274*(1611), 891–894; Hungry man preferences: Nelson, L. D., & Morrison, E. L. (2005). The symptoms of resource scarcity: Judgments of food and finances influence preferences for potential partners. *Psychological Science, 16*(2), 167–173.

42. Lap dancers: Miller, G., Tybur, J. M., & Jordan, B. D. (2007). Ovulatory cycle effects on tip earnings by lap dancers: Economic evidence for human estrus? *Evolution and Human Behavior, 28,* 375–381; Fertile days: Bryant, G. A., & Haselton, M. G. (2009). Vocal cues of ovulation in human females. *Biology Letters, 5*(1), 12–15.

43. Wedekind, C., Seebeck, T., Bettens, F., & Paepke, A. J. (1995). MHC-dependent mate preferences in humans. *Proceedings: Biological Sciences, 260*(1359), 245–249. Miller, S. L., & Maner, J. K. (2011). Ovulation as a male mating prime: Subtle signs of women's fertility influence men's mating cognition and behavior. *Journal of Personality and Social Psychology, 100*(2), 295–308.

44. Roberts, S. C., Gosling, L. M., Carter, V., & Petrie, M. (2008). MHC-correlated odor preferences in humans and the use of oral contraceptives. *Proceedings of the Royal Society B: Biological Sciences, 275*(1652), 2715–2722.

45. Earnings: Hamermesh, D., & Biddle, J. (1994). *American Economic Review,*84(5), 1174–1194; Doctors: Hadjistavropoulos, H. D., Ross, M. A., &

von Baeyer, C. L. (1990). Are physicians' ratings of pain affected by patients' personal attractiveness? *Social Science and Medicine, 31*(1), 69–72; Hiring: Ruffle, B. J., & Shtudiner, Z. (2010). Are good-looking people more employable? *Monaster Center for Economic Research, Working paper 10-06;* Attractiveness of talking: Borkenau, P., Maurer, N., Riemann, R., Spinath, F. M., & Angleitner, A. (2004). Thin slices of behavior as cues of personality and intelligence. *Journal of Personality and Social Psychology, 86*(4), 599–614.

46. Lomax, A. (1971). Choreometrics and ethnographic filmmaking. *Filmmakers Newsletter, 4,* 22–30; Critics of Lomax: Kealiinohomoku, J. W. (1991). Review essay: "The Longest Trail—film by Alan Lomax and Forrestine Paulay." *Yearbook for Traditional Music, 23,* 167–169. See also: Williams, D. (2007). On choreometrics. *Visual Anthropology, 20,* 233–239.

47. Richardson, D. C., Spivey, M. J., Edelman, S., & Naples, A. J. (2001). "Language is spatial": Experimental evidence for image schemas of concrete and abstract verbs. In *Proceedings of the Twenty-third Annual Meeting of the Cognitive Science Society* (pp. 873–878). Erlbaum. Meteyard, L. & Vigliocco, G. (2009). Verbs in space: Axis and direction of motion norms for 299 English verbs. *Behavior Research Methods, 41*(2), 565–574.

48. Maas, A., Pagani, D., & Berta, E. (2007). How beautiful is the goal and how violent is the fistfight? Spatial bias in the interpretation of human behavior. *Social Cognition, 25*(6), 833–852; Soccer fouls: Kranjec, A., Lehet, M., Bromberger, B., & Chatterjee, A. (2010). A sinister bias for calling fouls in soccer. *PLoS ONE, 5*(7), e11667.

49. Maass, A., & Russo, A. (2003). Directional bias in the mental representation of spatial events: Nature or culture? *Psychological Science, 14*(4), 296–301. People also tend to imagine time flowing in the same direction as their culture's writing: Fuhrman, O., & Boroditsky, L. (2010). Cross-cultural differences in mental representations of time: Evidence from an implicit nonlinguistic task. *Cognitive Science, 34*(8), 1430–1451.

50. Gilovich, T., Vallone, R., & Tversky, A. (1985). The hot hand in basketball: On the misperception of random sequences. *Cognitive Psychology, 17*(3), 295–314.

51. Gigerenzer, G., & Brighton, H. (2009). Homo heuristicus: Why biased minds make better inferences. *Topics in Cognitive Science, 1*(1), 107–143.

52. Rentfrow, P. J., & Gosling, S. D. (2007). The content and validity of music-genre stereotypes among college students. *Psychology of Music, 35*(2), 306–326.

53. Chamorro-Premuzic, T., & Furnham, A. (2007). Personality and music: Can traits explain how people use music in everyday life? *British Journal of Psychology, 98,* 175–185.

54. Peel, E. (1946). A new method for analyzing aesthetic preferences: Some theoretical considerations. *Psychometrika, 11*(2), 129–137.

55. Shermer, M. (1997). *Why people believe weird things: Pseudoscience, superstition, and other confusions of our time (p. 93).* W. H. Freeman.

56. McManus, M., & Davies, J. (under review). The effects of specific physical features on perceived intelligence.

57. Somel, M., Franz, H., Yan, Z., Lorenc, A., Guo, S., Giger, T., & Khaitovich, P. (2009). Transcriptional neoteny in the human brain. *Proceedings of the National Academy of Science, 106*(14), 5743–5748.

58. Trut, L. N. (1999). Early canid domestication: The farm-fox experiment. *American Scientist, 87,* 160–169.

59. The idea was presented by Richard Wrangham, quoted in: Taylor, J. (2009). *Not a chimp: The hunt to find genes that make us human*. Oxford University Press.

60. McAuliffe, K. (2010). The incredible shrinking brain. *Discover, 31*(7), 54–59.

61. Morey, D. F. (1992). Size, shape and development in the evolution of the domestic dog. *Journal of Archaeological Science, 19*(2), 181–204.

62. This also might be happening with bonobos (a great ape similar to a chimpanzee). McAuliffe, K. (2010). The incredible shrinking brain.

63. Judg.me Blog (2012). *What makes one appear smarter and more sociable?* Retrieved from http://judg.me/blog/judgment-day/ (May 22, 2012). See also the author photograph for this book.

64. Skolnick Weisberg, D., Keil, F. C., Goodstein, J., Rawson, E., & Gray, J. R. (2008). The seductive allure of neuroscience explanations. *Journal of Cognitive Neuroscience, 20*(3), 470–477.

65. Spina, R. R., Ji, L.-J., Guo, T., Zhang, Z., Li, Y., & Fabrigar, L. (2010). Cultural differences in the representativeness heuristic: Expecting a correspondence in magnitude between cause and effect. *Personality and Social Psychology Bulletin, 36*(5), 583–597.

66. Friedman, L. F. (2013). Outside in: But it's all natural! *Psychology Today,* 38.

67. Eidelman, S., Pattershall, J., & Crandall, C. S. (2010). Longer is better. *Experimental Social Psychology, 46*(6), 993–998.

68. Hood, B. M., & Bloom, P. (2008). Children prefer certain individuals over perfect duplicates. *Cognition, 106*(1), 455–462.

69. Begue, L., Charmoillaux, M., Cochet, J., Cury, C., & deSuremain, F. (2008). Altruistic behavior and the bidimensional just world belief. *American Journal of Psychology, 121*(1), 47–56.

70. Appel, M. (2008). Fictional narratives cultivate just-world beliefs. *Journal of Communication, 58*(1), 62–83.

71. Stegmueller, D., Scheepers, P., Rossteutscher, S., & de Jong, E. (2012). Support for redistribution in Western Europe: Assessing the role of religion. *European Sociological Review, 28*(4), 482–497. The belief in a just world can also make people feel empowered. Belief in a just world predicts earlier "coming out" in gay and bisexual men who, during childhood, had greater femininity. Presumably this is because it makes them feel less likely to be the victim of negative outcomes; see: Bogaert, A., & Hafer, C. (2009). Predicting the timing of coming out in gay and bisexual men from world beliefs, physical attractiveness, and childhood gender identity/role. *Journal of Applied Social Psychology, 39*(8), 1991–2019.

72. Bering, J. (2008). The end?: Why so many of us think our minds continue on after we die. *Scientific American Mind,* 34–31.

73. Bering, J. (2002). Intuitive conceptions of dead agents' minds: The natural foundations of afterlife beliefs as phenomenological boundary. *Journal of Cognition and Culture, 2*(4), 263–308.

74. Psychological continuity lessens with age: Bering, J. M., & Bjorklund, D. F. (2004). The natural emergence of afterlife reasoning as a developmental regularity. *Developmental Psychology, 40*(2), 217–233; Catholic kids hold on longer: Bering, J. M., Hernendez-Blasi, C., & Bjorklund, D. F. (2005). The development of "afterlife" beliefs in secularly and religiously schooled children. *British Journal of Developmental Psychology, 23*(4), 587–607; Wording can

affect psychological continuity: Harris, P.L., & Giménez, M. (2005). Children's acceptance of conflicting testimony: The case of death. *Journal of Cognition and Culture, 5*(1/2), 143–164.

7: WHY WE GET RIVETED

1. Dar-Nimrod, I., & Heine, S. J. (2011). Genetic essentialism: On the deceptive determinism of DNA. *Psychological Bulletin, 137*(5), 800–818.

2. Holyoak, K. J. (2005). *The Cambridge handbook of thinking and reasoning.* Cambridge University Press.

3. Dissanayake, E. (1995). *Homo aestheticus: Where art comes from and why.* University of Washington Press; Haidt, J. (2012). *The righteous mind: Why good people are divided by politics and peligion.* Pantheon Books.

4. Derrick, J. L., Gabriel, S., & Tippin, B. (2008). Parasocial relationships and self-discrepancies: Faux relationships have benefits for low self-esteem individuals. *Personal Relationships, 15*(2), 261–280.

5. Frazier, B. N., Gelman, S. A., Wilson, A., & Hood, B. M. (2009). Picasso paintings, moon rocks, and hand-written Beatles lyrics: Adults' evaluations of authentic objects. *Journal of Cognition and Culture, 9,* 1–14.

6. Bernhardt, P. C., Dabbs Jr., J. M., Fielden, Julie, A., & Lutter, C. D. (1998). Testosterone changes during vicarious experiences of winning and losing among fans at sporting events. *Physiology & Behavior, 65*(1), 59–62.

7. Pinker, S. (1994). *The language instinct: How the mind creates language.* HarperCollins.

8. For an eloquent description of this idea see Dawkins's TED talk "The Queer Universe."

9. If the solar system were scaled down so that the sun had the diameter of about half a meter, the earth would be 49 meters away and about 0.7 millimeters in diameter, about the size of a grain of sand.

10. This idea was eloquently described in: Atran, S. (2002). *In gods we trust: The evolutionary landscape of religion.* Oxford University Press. Chapter 3: God's creation: Evolutionary origins of the supernatural. See for an argument for why the predator-prey relationship is more dominant in religions than parent-child.

11. Peltzer, K. (2003). Magical thinking and paranormal beliefs among secondary and university students in South Africa. *Personality and Individual Differences, 35*(6), 1419–1426.

12. Atran, S. (2002). *In gods we trust.*

13. Dennett, D. C. (2006). *Breaking the spell: Religion as a natural phenomenon.* Penguin Books.

14. Atran, S. (2002). *In gods we trust.* Chapter 1: An evolutionary riddle.

15. "Then will they become Gods . . . they will never cease to increase and to multiply, worlds without end. When they receive their crowns, their dominions, they then will be prepared to frame earths like unto ours and to people them in the same manner as we have been brought forth by our parents, by our Father and God." Brigham Young, *Journal of Discourses* 17:143. Retrieved February 9, 2014 from: http://jod.mrm.org/18/257.

16. Harris, S., Kaplan, J. T., Curiel, A., Bookheimer, S. Y., Iacoboni, M., & Cohen, M. S. (2009). The neural correlates of religious and nonreligious belief. *PLoS ONE, 4*(10), e0007272.

17. Bloom, P. (2006). Is God an accident? In A. Gawande (Ed.), *The best American science writing 2006* (pp. 272–290). HarperCollins.

18. The Aztecs used mushrooms containing psilocybin. Religions north of the Aztecs used peyote, which contains mescaline. Religions south of the Aztecs used leaves and vines containing DMT (dimethyltriptamine). All these drugs are hallucinogens capable of inducing experiences described as "transformative" or "religious"; see: Haidt, J. (2012). *The righteous mind.* (p. 228).

19. Interpretation: Roberts, G., & Owen, J. (1988). The near-death experience. *The British Journal of Psychiatry, 153*(5), 607–617; Falling asleep: Moore, J. (1994). Moveable feasts: The Gurdjieff work. *Journal of Contemporary Religion, 9*(2), 11–16.

20. Rachman, S. (1997). A cognitive theory of obsessions. *Behavior Research and Therapy, 35*(9), 793–802. Erikson, E. H. (1958). *Young man Luther: A study in psychoanalysis and history.* W. W. Norton.

21. Whitehouse, H. (2000). *Arguments and icons: Divergent modes of religiosity* (pp. 130–131). Oxford University Press.

22. Landsborough, D. (1987). St Paul and temporal lobe epilepsy. *Journal of Neurology, Neurosurgery, and Psychiatry, 50*(6), 659–664.

23. Saver, J. L. & Rabin, J. (1997). The neural substrates of religious experience. *Journal of Neuropsychiatry, 9*(3), 498-510. Hamer, D. (2004). *The God Gene: How Faith is Hardwired Into Our Genes.* (p 132) Doubleday.

24. Radin, P. (1987). *The trickster: A study in American Indian mythology.* Schocken.

25. Whitehouse, H. (2000). *Arguments and icons.*

26. Boyer, P. (2001). *Religion explained: The evolutionary origins of religious thought* (pp. 283–285). Basic Books.

27. Ibid. (p. 69).

28. Iannaccone, L. R. (1991). The consequences of religious market structure: Adam Smith and the economics of religion. *Rationality and Society, 3*(2), 156–177.

29. Dennett, D. C. (2006). *Breaking the spell.*

30. Ibid.

31. Stanovich, K. E. (2004). *The robot's rebellion.* University of Chicago Press.

32. Dennett, D. C. (2006). *Breaking the spell.*

33. Stanovich, K. E. (2004). *The robot's rebellion.*

34. Sanderson, S. K., & Roberts, W. W. (2008). The evolutionary forms of the religious life: A cross-cultural, quantitative analysis. *American Anthropologist, 110*(4) 454–466.

35. Atran, S. (2002). *In gods we trust.* Chapter 1: Introduction: An evolutionary riddle.

36. Boyer, P. (2001). *Religion explained.*

37. I make exceptions for scientists who are working at the cutting edge of their fields. Then they may disagree with the scientific consensus. If no scientists disagreed with the consensus, science would never change! But I do think it is intellectually unjustified to disagree with current scientific findings for those people who are not well versed in the particular scientific field making the claim. So, for example, if you don't know a lot about physics, you are not intellectually justified in disagreeing with accepted, textbook-level knowledge in physics.

38. Haidt, J. (2012). *The righteous mind.*

39. Religious charity: Johnson, B. (2012). Religion and philanthropy. Unpublished manuscript, University of North Carolina at Chapel Hill; Socially strategic: Pyrzycki, B. G., Finkel, D. N., Shaver, J., Wales, N., Cohen, A. B., & Sosis, R. (2012). What does God know? Supernatural agents' access to socially strategic and non-strategic information. *Cognitive Science, 36*(5), 846–869;

Sacrifice for group: Atran, S. (2002). *In gods we trust*. Chapter 1: Introduction: An evolutionary riddle.

40. Frejka, T., & Westhoff, C. F. (2008). Religion, religiousness, and fertility in the US and Europe. *European Journal of Population, 24*(1), 5–31. Sosis, R. (2000). Religion and intragroup cooperation: Preliminary results of a comparative analysis of utopian communities. *Cross-Cultural Research, 34*(1), 70–87. Laboratory and field studies show that when people calculate costs and benefits, it leads to a breakdown in societal common resources; see: Atran, S. (2002). *In gods we trust*. Chapter 8: Culture without mind: Sociobiology and group selection.

41. Haidt, J. (2012). *The righteous mind*. (p. 257).

42. Thinking about science makes you moral: Ma-Kellams, C. & Blascovich, J. (2013). Does "science" make you moral? The effects of priming science on moral judgments and behavior. *PLOS ONE, 8*(3), e57989. For an extended argument, with evidence, see: Haidt, J. (2012). *The righteous mind*.

43. Magical thinking: Hutson, M. (2008). Magical thinking. *Psychology Today, 41*(2), 88–95; Pleasure: Mohr, C., Landis, T., Bracha, H. S., Fathi, M., & Brugger, P. (2005). Levodopa reverses gait asymmetries related to anhedonia and magical ideation. *European Archives of Psychiatry and Clinical Neuroscience, 255,* 33–39; Swimmers: Starek, J. E., & Keating, C. F. (1991). Self-deception and its relationship to success in competition. *Basic and Applied Social Psychology, 12*(2), 145–155.

44. Trees: Atran, S., Medin, D., Ross, N., Lynch, E., Vapnarsky, V., Ucan Ek', E., Coley, J., Timura, C., & Baran, M. (2002). Folkecology, cultural epidemiology, and the spirit of the commons: A garden experiment in the Mayan lowlands, 1991–2001. *Current Anthropology, 43*(3), 421–450; Good luck charms: Damisch, L., Stoberock, B., & Mussweiler, T. (2010). Keep your fingers crossed! How superstition improves performance. *Psychological Research, 21,* 1014–1020; Panic: Inzlicht, M., McGregor, I., Hirsh, J. B., & Nash, K. (2009). Neural markers of religious conviction. *Psychological Science, 20*(3), 385–392; Negative emotions: Sharp, S. (2010). How does prayer help manage emotions? *Social Psychology Quarterly, 73*(4), 417–437; Blood pressure: Sorensen, T., Danbolt, L. J., Lien, L., Koenig, H. G., & Holmen, J. (2011). The relationship between religious attendance and blood pressure: The HUNT study, Norway. *International Journal of Psychiatry in Medicine, 42*(1), 13–28; Stress: Belding, J., Howard, M., McGuire, A., Schwartz, A., & Wilson, J. (2010). Social buffering by God: Prayer and measures of stress. *Journal of Religion and Health, 49*(2), 179–187; Dopamine: Hamer, D. (2004). *The God gene: How faith is hardwired into our genes* (p. 165). Doubleday.

45. Mormons: Enstrom, J., & Breslow, L. (2008). Lifestyle and reduced mortality among active California Mormons, 1980–2004. *Preventive Medicine, 46*(2), 133–136; Attending church: Chida, Y., Steptoe, A., & Powell, L. H. (2009). Religiosity/spirituality and mortality. *Psychotherapy and Psychosomatics, 78*(2), 81–90; HIV: Finocchario-Kessler, S., Catley, D., Berkley-Patton, J., Gerkovich, M., Williams, K., Banderas, J., & Goggin, K. (2011). Baseline predictors of ninety percent or higher antiretroviral therapy adherence in a diverse urban sample: The role of patient autonomy and fatalistic religious beliefs. *AIDS Patient Care & STDs, 25*(2), 103–111; Education: Moulton, B. E., & Sherkat, D. E. (2012). Specifying the effects of religious participation and educational attainment on mortality risk for U.S. adults. *Sociological Spectrum, 32*(1), 1–19.

46. Vaccinated: Ruijs, W. L. M., Hautvast, J. L. A., van der Helden, K., de Vos, S., Knippenberg, H., & Hulscher, M. E. J. L. (2011). Religious subgroups influencing vaccination coverage in the Dutch Bible belt: An ecological study. *BMC Public Health, 11*(102); Past life belief: Meyersburg, C. A., & McNally, R. J. (2011). Reduced death distress and greater meaning in life among individuals reporting past life memory. *Personality and Individual Differences, 50*(8), 1218–1221; Anxiety: Inzlicht, M., McGregor, I., Hirsh, J. B., & Nash, K. (2009). Neural markers of religious conviction. *Psychological Science, 20*(3), 385–392; Hearing about religion: Inzlicht, M. & Tullett, A. M. (2010). Reflecting on God: Religious primes can reduce neurophysiological response to errors. *Psychological Science, 21*(8), 1184–1190; Happier: Steger, M. F., & Frazier, P. (2005). Meaning in life: One link in the chain from religiousness to well-being. *Journal of Counseling Psychology, 52*(4), 574–582; Belief certainty: Galen, L. W., & Kloet, J. D. (2010). Mental well-being in the religious and the non-religious: Evidence for a curvilinear relationship. *Mental Health, Religion, and Culture,* (iFirst), 1–17; Agnostics: Morchon, D., Norton, M. I., & Ariely, D. (2011). Who benefits from religion? *Social Indicators Research, 101,* 1–15; Religion valued: Diener, E., Tay, L., & Myers, D. G. (2011). The religion paradox: If religion makes people happy, why are so many dropping out? *Journal of Personality and Social Psychology, 101*(6), 1278–1290.

47. Atran, S. (2002). *In gods we trust.* Chapter 1: Introduction: An evolutionary riddle.

48. This interesting theory has not, to my knowledge, been directly tested. It is described in: Atran, S. (2002). *In gods we trust.* Chapter 1: Introduction: An evolutionary riddle, and Chapter 4: Counterintuitive worlds: The mostly mundane nature of religious beliefs.

49. Epley, N., Converse, B. A., Delbosc, A., Monteleone, G. A., & Cacioppo, J. T. (2009). Believer's estimates of God's beliefs are more egocentric than estimates of other people's beliefs. *Proceedings of the National Academy of Sciences of the United States of America, 106*(51), 21533–21538.

INDEX